P9-DMA-711

THE SURVIVORS CLUB

Bob -

Take a lickin',

Keep on tickin'!

David + Terry

5-23-09

ALSO BY BEN SHERWOOD

FICTION

The Man Who Ate the 747

The Death and Life of Charlie St. Cloud

THE SURVIVORS CLUB

The Secrets and Science that Could Save Your Life

BEN SHERWOOD

GRAND CENTRAL
PUBLISHING

New York Boston

The content in *The Survivors Club* book and Survivor Profiler is for informational purposes only and not intended as a substitute for professional advice, diagnosis, or treatment. Always consult your physician or other qualified expert with questions regarding a specific condition or challenge. Never disregard professional advice or delay seeking help because of content you have found in *The Survivors Club* book or your Profiler Report. Reliance on any information provided in *The Survivors Club* book or Survivor Profiler is solely at the reader's discretion and risk.

Grand Central Pubishing
Hachette Book Group
237 Park Avenue
New York, NY 10017

Visit our Web site at www.HachetteBookGroup.com.

Printed in the United States of America

First Edition: January 2009
10 9 8 7 6 5 4 3 2 1

Grand Central Publishing is a division of Hachette Book Group, Inc.
The Grand Central Publishing name and logo is a trademark of Hachette Book Group, Inc.

Library of Congress Cataloging-in-Publication Data

Sherwood, Ben.
The survivors club : the secrets and science that could save your life / Ben Sherwood.
p. cm.
ISBN: 978-0-446-58024-3 (regular edition)
ISBN: 978-0-446-54123-7 (large print edition)
1. Survival skills. 2. Accidents. 3. Medical emergencies. I. Title.
GF86.S538 2009
613.6'9—dc22

2008016203

Book design and text composition by L&G McRee

To William Richard Sherwood

CONTENTS

PROLOGUE

Brace for Impact

First they tell you not to panic and then they try to drown you.

Honeyed light settles over the US Marine Corps air station in Miramar, California, just a few miles from the Pacific Ocean. A warm breeze ruffles the palm trees. But the golden surroundings are deceptive. The military sends its sailors here to learn the art and craft of escaping from crashing jets and sinking helicopters. If your job involves flying over water—and that pretty much includes everyone in the navy—you come to the Aviation Survival Training Center to learn how to survive a "mishap," the euphemism for an accident or worse. There are four men and one woman in my class, and right from the start in our "welcome aboard" briefing, we're told this training is "high risk." An instructor hands out a government form to sign. It's a release discharging the United States from any and all claims of injuries.

The navy takes men and women barely out of high school and teaches them to survive dangerous missions in hostile environments. It prepares people with no prior experience to endure the unimaginable. It molds warriors *and* survivors. That's why my research begins here behind barbed wire and guard posts where they don't usually invite civilians. I want to experience the stress of survival training and immerse myself in the military's culture and customs of staying alive. Above all, I want to learn the secrets of who survives.

We spend the first eight hours in a classroom with intense, highly focused experts who hit us rapid-fire with information about every imaginable survival threat. They rattle off the different kinds of hypoxia—not enough oxygen—and how to recognize the symptoms. They plow through sensory physiology—how our senses often trick us into making deadly mistakes. They emphasize the importance of situational awareness—S/A—which means knowing what's going on around you at any given moment and being able to anticipate danger. They also instruct us in some very practical matters: No matter how thirsty you are, never drink urine and *never ever* drink seawater.* And finally, they insist that the key to survival is attitude. "If you lose that will to live," one burly instructor says, "odds are you aren't going to make it." By late afternoon, we're finished in the classroom. Now it's time for the real reason we're here: dunking.

I'm standing in my trunks on the pool deck, sucking in my stomach and checking out the rest of the class. They're all ridiculously lean and fit. Right next to me: a ramrod young man who received the Sailor of the Quarter award from his command. Next to him: the crisp and wiry commanding officer of a destroyer. I figure I don't stand much of a chance. For the opening test—can you swim?—here's what they do. First, they dress you up in full flight gear: helmet, flight suit, steel-toed boots, anti-g pants, parachute harness, survival vest, life vest, and gloves. Then they throw you in the pool and make you tread water till your heart is about to explode. Without a break, they force you to do something

* In a British study of 163 "life craft voyages," 38.8 percent of the people who drank seawater died compared with 3.3 percent who did not. Seawater consists of 3 percent dissolved salt, and drinking it—even diluted—"aggravates dehydration" and "hastens death." Most survival experts also advise against drinking urine because of the salt content, as well as waste products that can make you sick.

called drown-proofing, a vexing technique of floating like a dead man in order to conserve energy and catch your breath. Many just sink and fail the course. If you make it through the preliminary tests, they've got a kind of extreme water park waiting for you. They whip you across a big pool with pulleys and wires to simulate a parachute dragging you over the ocean. They yank you up a cable while drenching you with the most powerful showerheads you've ever seen to imitate the rotor blast of a rescue helicopter. And finally, the grand finale: They strap you into a chair, crash you into the pool, flip you upside down, sink you to the bottom, and tell you to escape.

I'm six foot four, so the first challenge is to find flight gear that's big enough. When I'm finally suited up, complete with size XL helmet, I can't help thinking of doomed presidential candidate Michael Dukakis and how ridiculous he looked on that M1A1 tank in 1988. I just hope I'm invisible as I submerge in the soothing eighty-five-degree water. We're in the shallow end, and I'm wearing twenty-five pounds of equipment. My feet slosh in size 13 boots that feel like clown shoes. I try to figure out the most efficient way to move, let alone swim. It's clear that the harder I try, the less effective I am. I come up with improvised mini strokes—a modified dog paddle—that seem to work pretty well. I look around at the rest of the class, relaxed and bobbing in the pool, and I get a competitive rush. I'm a little jittery—my breathing is shallow—and I don't want to fail. This trepidation is exactly what they're trying to foment. Until you're in the water, trying to float with all that gear, you can't even begin to imagine what it's like in the open ocean fighting for your life. In this safe and controlled environment, I'm experiencing the first sparks of the survival instinct: brain chemicals flowing, competitiveness surging, all thoughts focused on a plan to stay afloat.

The whistle blows, and we're supposed to swim across

the twenty-five-yard pool. I go all out with my improvised mini strokes. I win the sprint, but I've expended too much energy. Now I'm worried I won't get past the next challenge—treading water and drown-proofing. Again, my brain locks on to a strategy for conserving strength: I alternate between dog paddling and floating on my back using the helmet for buoyancy. When the whistle sounds at the end of this endurance trial, two of my classmates—the ramrod sailor and a female photographer from an aircraft carrier—have been eliminated. They flunked because they sank. Eight percent of the folks who come here fail the first time.

The woman who runs the survival school is waiting by the pool when I drag myself from the water. Lieutenant Commander Rebecca Bates is five foot eleven with short blond hair streaked by sun and chlorine. She's lean, terse, and wears a crisp green flight suit. Her radio call sign is "Sparky" but behind her back the staff refers to her as "Big Momma." As I stand dripping in front of her, she seems surprised, even a little amused, that I passed the drown-proofing test. "Are you made of cork?" she jokes. Then she lets me in on a secret: The instructors check out the candidates when they emerge in swimsuits from the locker room. Some get an immediate *oh no* reaction because they're destined to fail. When she first saw me, she thought I didn't have a chance.

Now it's time for the Devices, each with its own convoluted military acronym. These are ingenious contraptions designed to simulate every underwater survival scenario. First: the SWIMMER. That's short for—ready?—Shallow Water Initial Memory Mechanical Exit Release trainer. It's an underwater door with half a dozen different handles that you might find on navy aircraft. You take a breath, submerge, and swim a certain length to the door, open various levers and latches, then push your way through. If you succeed, you get to do it again with blackout goggles.

Next: the SWET. It stands for Shallow Water Egress Trainer. Imagine a little tower with a chair on top sticking out of the shallow end of the pool. You are strapped into the seat perched above the water, and then someone pulls down on a lever, flipping the chair (and you) upside down into the pool. It's like a human rotisserie. With a great splash, you're submerged and inverted. The challenge: Unbuckle your harness and swim out. If you make it, they fasten you into the chair again and you do it blindfolded.

Lieutenant Commander Bates is still smiling as I emerge again from the pool. She can't quite believe I made it through the SWIMMER challenge and the SWET spin cycle, but now she's got one final obstacle called the 9D6. I am led to a nearby building with a big indoor pool. The giant jib arm of a crane hangs over the water with a big blue gondola dangling on thick cables. It's the size and shape of a UPS truck. This is the METS, short for Multi-place Underwater Egress Trainer. The folks here call it the Dunker, a machine that simulates a helicopter crashing in the ocean. It's engineered to plunge into the pool, roll over, and sink upside down just like a real chopper that's top-heavy with its engine and rotors.*

This is the grand finale of water survival training. I ease into the pool, and the crane lowers the sixty-four-hundred-pound Dunker. I notice bubbles rising from the depths. Two navy divers are already waiting for me at the bottom. They're the rescue team if I get in trouble. Another trainer—called a safety swimmer—helps me into the main cabin, which is set up to look like a typical transport helicopter with six crew seats. Everything starts to shift and swing as the giant jib hauls us up into the air. The trainer leads me to the front of

* On the day of my visit to the Naval Survival Training Institute in Pensacola, Florida—the national headquarters of the navy's survival programs—a Seahawk helicopter crashed off the coast of San Diego, killing four sailors on board. Just nine seconds after the first sign of trouble, the helicopter hit the ocean and rolled on its right side. The Seahawk was recovered thirty-seven hundred feet below the surface. Three of the four bodies were still inside.

the gondola, where he buckles me into the pilot's seat. Between my legs, I've got a fake cyclic control stick for steering. My feet rest on pretend pedals to direct the angle of the tail rotor blades.

I hear the crane whir and the cables creak. In the final seconds before dunking, I recall the most important lesson of my classroom training. If the experience gets too intense or if I need help at any point, I'm supposed to use a simple, unmistakable rescue signal. The exercise will stop immediately and the frogmen will pull me to safety. That sign couldn't be easier or more apt. I'm supposed to press my palms together in front of my face and pray.

"Brace for water impact," the instructor shouts. The cockpit shudders, the Dunker drops, and we hit the pool. Water surges through the floor and windows, and I try to focus on my goal: timing it just right so I inhale one big gulp of air before going under. But the cabin suddenly lurches and plunges and it feels like I'm being sucked down a giant drain. I barely capture a mouthful of oxygen, let alone a deep breath. I curse to myself. As the water whooshes to the ceiling and the cabin spins and sinks, the twirling and churning sensation feels thrilling, but then I see only bubbles. The world goes white. The big blue gondola settles at the bottom of the pool. Upside down, fifteen feet underwater, and buckled into a metal canister, I'm completely disoriented. The first twinges of alarm ripple through my brain: *How long will I last? Can I get out? Will I survive?*

INTRODUCTION

The Survivors Club

The field is known as "human factors in survival." Translation: Why do some people live and others die? How do certain people make it through the most difficult trials while others don't? Why do a few stay calm and collected under extreme pressure when others panic and unravel? How do some bounce back from adversity while others collapse and surrender?

This book answers those questions. It shares the true stories of regular people who have been profoundly tested by life—men and women who have been beaten down, sometimes literally flattened. It explores how ordinary folks somehow manage to pick themselves up, again and again, in the face of overwhelming odds. It investigates whether survivors are different from you and me. And it dissects the mind-set and habits that are shared by the most effective survivors. In short, it unlocks the secrets of who lives and who dies and shows how you can improve your chances in virtually any crisis.

At the outset, I'd like to put a few things on the table. Almost everyone I know has faced—or is coping with—some kind of serious challenge or adversity. I wrote this book for them and for myself. While I certainly haven't been tested like the survivors in these pages, I've hit some bumps and experienced my share of loss and grief. My father was in excellent health when he died suddenly at age sixty-four from a massive and inexplicable brain bleed. Defying the probabilities, my mother has beaten back ovarian cancer for nine years, always deflecting credit to the aggressive treatment orchestrated by her superb oncologist. As a journalist, I've had a few close scrapes and witnessed plenty of tragedy. In August 1992, while covering the bloody siege of Sarajevo for ABC News, I was sitting shoulder-to-shoulder with a veteran producer and friend named David Kaplan when he was fatally wounded by a sniper. A nine-millimeter bullet ripped through the back door of our Volkswagen van, pierced David's back, and severed his pulmonary artery. French combat surgeons fought to save him, but his injuries were too grave. It was pure chance that he—not I—ended up in that fatal middle seat, which had seemed the safest spot, away from the windows.

I've always been something of a control freak, so each of these events called everything into question. Why do healthy people drop dead without reason? How can cancer strike those who aren't at risk? Why do bullets find one victim and not another? Perhaps in an attempt to regain some command, I began to ask: Are there any hidden ways to improve the odds? If "no one here gets out alive," as Jim Morrison sings, what are the tricks of sticking around as long as possible? My search produced this book, and the answers are both humbling and comforting. When it comes to survival, as you'll see, there's a whole lot that you can't control, and a surprising amount that you can.

A few other disclaimers: I'm not a survivalist or an outdoorsman. I don't stockpile canned goods and I'm not preparing for Armageddon, although I did buy emergency kits for my car and home while researching this book. I'm a city guy, a journalist, and an occasional novelist. I've spent most of my life asking questions and I've always been drawn to stories of people under pressure. I remember the summer at age ten when I began to read

Alive, the astonishing saga of a plane crash in the Andes Mountains and the passengers who endured seventy-two freezing days on a glacier. It's human nature to speculate: What would I have done? Would I have pushed myself to the same extremes? In March 2000, while working for *NBC Nightly News,* I marveled at images of Sofia Xerindza, a woman in Mozambique who escaped the deadly floodwaters of the Limpopo River by climbing into a tree, where she gave birth to a baby girl.*

At ABC's *Good Morning America,* where I worked as executive producer for two and a half years, I watched a veritable parade of survivors on the screen and always wondered: How did these people endure their trials? Were they always so strong and resilient, or did these abilities suddenly materialize when they most needed them? Television interviews last only a few minutes, so what would these survivors say if the clock wasn't ticking? How did they *really* get through it? In quieter moments, what wisdom might they share about their experiences? I also wanted to know about all the people who face life's everyday challenges without any attention or fanfare, the unheralded folks fighting illnesses like cancer, Alzheimer's, and Parkinson's. In the face of life's inevitable crises, how do they get through their days? Where do they find the fortitude, sometimes literally, to climb out of bed? And selfishly, how could I get some of their strength?

In this book you'll meet survivors of every imaginable ordeal, young and old, rich and poor, the guy down the street and people in the news. I've gathered tales on every continent; if you can conceive of a crisis, I've probably interviewed someone who has gone through it and come out on the other side. A woman doused with gasoline and set on fire by her husband; a bicyclist on a morning ride crushed by a twenty-one-ton truck;

* For photographs and videos of survivors described in this book, please visit www.TheSurvivorsClub.org.

veterans who lived through the Japanese sneak attack on Pearl Harbor and survived the great battles of World War II; a young ballerina forced to dance for her life by Dr. Josef Mengele in Auschwitz.

The Survivors Club will explain how they did it. You'll learn some of the secrets of survival—like the safest seat on an airplane, the best place to suffer a heart attack, and how the number 3 could keep you alive in a crunch. You'll discover how some people are born with a Resilience Gene that actually protects them from the worst knocks in life. You'll find out how a few easy changes in your food and vitamins can boost your ability to bounce back from hardship. In Manitoba, Canada, you'll meet the human Popsicle: a professor who has dunked himself in ice thirty-nine times in order to understand freezing to death. In Boston, Massachusetts, you'll sit down with the Harvard Medical School expert who specializes in cases of people who are literally scared to death. And in England, a magician-turned-psychology-professor will welcome you to Luck School and show you how to increase your good fortune by 40 percent. Perhaps most surprising of all, in Charlotte, North Carolina, you'll discover the emerging field of posttraumatic growth and the remarkable theory that more people benefit from life's worst events than are shattered by them.

The two questions at the heart of this book are these: (1) *What does it really take to survive?* And (2) *What kind of survivor are you?* The answers will unfold in two sections. In part 1, I'll investigate the keys to survival in everyday crises ranging from car wrecks to violent crimes. I'll take you inside one of the country's top hospitals to explore who lives and dies in emergency rooms and why, for instance, the ideal age for a brain injury is around sixteen. I'll delve into the psychology of survival and what specific personality traits give you the greatest advantage in beating the odds. I'll explore whether the will to live makes a difference in defeating diseases like breast cancer. And I'll take you on a pilgrimage to a little chapel in the Sangre de Cristo Mountains of northern New Mexico where they believe that miracle cures really happen.

At every step, you'll encounter the wisdom of men and women who have fallen into the abyss and somehow climbed out. These survivors want to share the tactics and strategies they wish they had known *before* their ordeals. In the race to survive, their insights may give you a critical head start. At the end of each chapter, I'll also try to unlock some of the mysteries of survival. For instance, how can a 145-pound grandmother lift a 3,450-pound Chevy Impala off her son? How did a French woman who smoked cigarettes and ate chocolate every day manage to survive to the age of 122? Why do right-handers live longer than lefties? Why are birthdays and holidays especially dangerous to your survival? And what can we learn from a five-thousand-year-old pine tree named Methuselah, perhaps the oldest living thing on earth?

Part 2 of this book shifts the focus to *you*. Are you as resilient as Trisha Meili, the Central Park Jogger raped, beaten, and left in a pool of blood? Are you as tenacious and tough as John McCain, tortured as a POW for five and a half years in Vietnam? Are you even remotely as competitive as cycling champ and cancer survivor Lance Armstrong? Are you as optimistic as Michael J. Fox, afflicted with Parkinson's disease since 1991? Working with a group of top psychologists and experts, I've developed a powerful new Internet tool to help you figure out your survivor personality. Analyzing the traits of more than one million people, my team has identified the five main types of survivors and the twelve most critical survival tools. With our exclusive test, you'll be able to find out your Survivor IQ. It will only take a few minutes, and when you're done, you'll get a customized report spelling out your unique survivor strengths. At any point, you can skip ahead to part 2 on page 293 and learn how to take the test. The Profiler will also produce your Survivor Match, which compares your personality with the men and women you're reading about in these pages. The readiness is all, Hamlet says, and I sincerely believe the Profiler can give you an edge when adversity strikes.

The blast of water up my nose scrambles everything.

I'm strapped upside down and can't see a thing at the bottom of the fuzzy chlorinated pool. That's nothing compared with the gurgling in my nostrils and the burning in the back of my throat. My concentration is shot. In my frazzled head, some kind of hazard light switches on. I try to focus on my escape plan and jam my gloved fingers into the snaps on the harness holding me in the Dunker. I shake them furiously and push away from the seat but don't get very far. The straps bite into my shoulders. I'm stuck.

I can feel the first tug in my lungs. I'm very conscious of the ensuing alarm and monologue in my brain. *Oxygen, please! You're underwater. You don't have all day. What are you going to do now?* I pry my fingers into the release latches again and shake them around. Then I yank hard on the harnesses and kick against the floor. I'm still trapped. I look down at my chest to see if I can figure out the problem. No chance. There are too many bands and clasps on my gear. There are also too many bubbles. Now, as my lungs start to clench, the warning sirens start clanging in my head. My brain is no longer asking politely. It's shouting: *Oxygen, now!*

In the classroom a few hours earlier, the instructors insisted that virtually anyone can be taught to get out of the Dunker. Their mantras were succinct. First, *maintain your reference point*. In the most chaotic situations, that means identifying and holding on to something that will help you stay oriented no matter how many times you flip over or get banged in the head. If you keep your point of reference, you will never get lost or confused and will always find a way out. I chose the handle on a door as my reference point. They didn't warn me the shot of water up my nose would distract me and make me let go. Still: That's the whole point. They're trying to disorient you. They want you to fail before you succeed. Their second mantra is to *wait for all sudden and violent motion to stop*. In the Dunker, that means surrendering to the whirling sensation of what it must feel like in a blender. Eventually, the chaos subsides, the tumult ceases, and it's a lot easier to handle the situation. Underwater, it can feel like an eternity, but it's typically only ten or fifteen seconds. That means you have plenty of air and time to unbuckle and get out. It all sounded great in theory.

Another pang in my lungs. I'm really starting to worry. I try one more time with the buckles, poking, jiggling, and finally thrashing with my fingers. No luck. Now, slamming hard into the floor with my legs, I throw myself against the harness with all my weight. This is my version of brute force. It's my last gasp. If this doesn't work, I'll resort to the prayer sign. I strain and flail and then—incredibly—I'm free, floating in the cockpit. I'm so surprised and elated that I actually marvel for a moment at my weightlessness. I drift around, unfettered, but quickly realize that I'm lost and don't know the way to the designated exit hatch. Even more confusing, I'm not sure which way is up. I swish in every direction trying to identify the right door. Then I find it, fumble with the latch, and force my way out. I follow the bubbles up toward the light. With my lungs straining, I break the surface and take a huge gulp of air.

The sun is setting over the palm trees at Miramar. F-18s thunder into the orange dusk. After thirteen hours of survival training, class is over, and our soaking flight suits are lined up on racks. Our boots, leaking great puddles, are set out to dry. As I get ready to go, the divers who watched from the bottom of the pool tell me that I looked a little "frenzied" trying to get out of the Dunker. Easy for them to say. Turns out my harness had actually released the first time I unbuckled it, but my survival vest had gotten snagged on the pilot's seat. In my turmoil, I hadn't been able to diagnose the problem. The frogmen say I had plenty of air and time and should have stayed calmer under pressure. Panic is the enemy. All those mental hazard lights, alarms, and sirens short-circuit our problem solving. We forget our training. We don't maintain our points of reference. We don't wait for the violent action to stop. We lose our minds and our way.

No matter the adversity, the navy says, survival is a mentality, a way of thinking. Survival is also a lens, a way of perceiving the world around you. The best survivors in the military share a constant outlook and approach, which they believe can also be applied to the struggles of everyday life. They understand that crisis is inevitable and they anticipate adversity. When they face a challenge, they observe and analyze the situation, devise a plan, and move decisively. If things go wrong, they adapt and improvise. If they get overwhelmed, they recover quickly. They also know how to wait for the worst to end. Understanding that even misfortune gets tired and needs a break, they're able to hold back, identify the right moment, and then do what they need to do. Psychologists have a clunky term for this: *active passiveness*. It means recognizing when to stop and when to go. In a critical sense, doing nothing can mean doing something. Inaction can be action, and embracing this paradox can save your life.

Lieutenant Commander Bates meets me at the front doors of her training center and hands over my exit papers or "qual sheets." She tells me I've earned a Q, the coveted grade for "qualified." Two other members of my class also receive their Qs. "Congratulations," she says with a farewell handshake. "You survived."

PART I

What It Takes to Survive

Whoever survives a test, whatever it may be, must tell the story. That is his duty.

—Elie Wiesel

1

A Knitting Needle Through the Heart

The Three Rules of the Survivors Club

The knitting needle pierced her heart. Then it saved her life.

Ellin Klor savors the irony, but it wasn't always so, especially when doctors cracked open her chest in the operating room to pry out the wooden needle that had punctured her breastbone and penetrated her right ventricle. Today her fingers stroke her yellow blouse, tracing the spot where she was speared. You would think that a spike as thick as a Number 2 pencil in your heart would finish you off. But, no.

January 9, 2006, was her lucky day.

It began as an ordinary Monday. The fifty-six-year-old children's librarian went to work in Santa Clara, California, then drove the after-school car pool for her daughter and fixed dinner for her family. Klor is a spark plug, pulsing with the energy of countless hobbies and an endless list of projects. Her choice that night was whether to make table decorations for a scholarship fund-raiser at school or to go to a meeting of her new knitting group. She almost stayed home but was anxious to show the gang some new patterns. So she grabbed three bags stuffed with books, yarn, and needles, and headed to a friend's house on Portal Avenue in Palo Alto. The knitting circle had been meeting for less than a year, and Klor loved being the teacher.

She parked her tan station wagon on the quiet street lined with London plane trees. Already late, she could tell from the other cars that some of the knitters had arrived. She hoisted her bags from the backseat. "The scourge of a librarian," she recalls, "carrying too much stuff around." Hurrying across the sidewalk, she followed the pathway to the one-story ranch house. The curtains were drawn and the porch was lit softly. Klor climbed the first of two wide steps, hardly treacherous, and then stubbed her foot. Suddenly she was falling down. Hands full with three bags, she tumbled forward and slammed into the ground, landing chest-first on a sack filled with unfinished knitting. She rolled over, stood right up, and scolded herself: *You shouldn't have been carrying so many things.*

Klor is five foot four with soft hazel eyes and a generous round face. She's admittedly a little plump and has always been a bit of a klutz, banging into things and tipping over, so her latest spill wasn't exactly a surprise. A quick check: Her knee was scraped but her clothes weren't torn. When she took a breath, her chest hurt, but she figured it was nothing. So she collected herself, gathered up the bags, knocked on the door, and was greeted by her girlfriend.

Inside, the knitters were already working in the living room. Klor wanted to get started, but the ache in the middle of her chest was growing worse with each breath. It wasn't an ordinary pang. This was different. She looked down at her red Façonnable sweater and lifted it up. The next image is ingrained in her memory. A jagged splinter of a wooden knitting needle, nearly four inches long, was jutting from her chest. It had broken in half, piercing her clothing and lodging in the middle of her bra right between her breasts.

"Oh my God," she whispered.

Her friends gasped at the needle and urgently calculated the options. First and foremost, should they try to pull it out? "No, don't touch it," Klor declared. It was pure instinct: She didn't want anyone to go near the injury until she was at the hospital. Doctors would say later this was the first decision that helped save her life. Plucking the spike would have been like pulling a plug or uncorking a bottle, and she might have bled out in the living room. Indeed, when Australian crocodile hunter Steve Irwin was speared in the chest by a bull ray while snorkeling in the Great Barrier Reef in 2006, some experts believe his fatal mistake was yanking out the stinger. The ray's venom didn't kill him. Rather, when he ripped the serrated barb from his chest, it wounded his left atrium and ventricle, causing more bleeding and cardiac arrest. The damage from pulling out the stinger was far greater than the trauma of it going in.

Now Klor and her friends faced the next critical question: Should they jump in a car and race to the emergency room? "No," Klor decided. "Call 911 right now." Waiting for the paramedics was a second lifesaving choice. If the needle had moved even the slightest amount in transit to the ER, the injury to her heart might have proven fatal. So Klor carefully sat down on a sofa to wait for the ambulance. She felt alert and even noticed something very odd. She had been impaled and yet there wasn't a single drop of blood anywhere. How was this possible? The next string of images flew by like a strange TV drama about herself. Paramedics. Stretcher. Sirens. IV. Oxygen. Emergency room. CT scan.

At the Stanford University Medical Center in Palo Alto, Klor waited anxiously for the ER doctors to tell her the extent of her problems. To distract herself, she focused on her daughter, Callie. Klor had waited until she was forty-two to have a child and had been blessed with a beautiful girl. Ever since Callie's birth in April 1993, Klor had found real joy in life. Now in the ER with a knitting needle jutting from her chest, she wondered: *How can I die when I'm finally happy?* The answer was clear: She needed to stay alive for Callie. Her thoughts also turned to her husband, Hal, a research engineer and tough guy who once hiked two miles on a broken ankle. His idea of a vacation involved trekking in the Himalayas, and he sometimes teased her lovingly that she was "a little wimpy." *What would Hal say when he heard about this?*

When the ER team finally briefed her on the results of her scans, she felt the first flood of fear. Their tone was urgent. The needle had penetrated her sternum, the long flat breastbone that's supposed to protect the heart, lungs, and major blood vessels from trauma. Over the years, her physicians had extracted every imaginable object sticking from every conceivable body part, but they told her this was brand-new. With fifty million knitters in the United States, there were literally hundreds of millions of needles across the country, but in the trauma world Ellin Klor was a celebrity. Paparazzi-style, a young doctor snapped her photo and then took mug-shot close-ups of the offending needle. Then the doctors delivered the really scary news. The point of the needle had grazed her heart, nicking the right ventricle. They could see internal bleeding. They needed to operate as soon as possible. Klor gave them her consent, and they rolled her up to the surgical suite and prepped her for the operation. This was her last memory of the ordeal.

Less than an hour after her tumble on the porch stairs, trauma surgeons would cut her chest open and crack her sternum. They would stitch up her heart. They would wire her breastbone back together and sew her up. They would leave a seven-inch scar from her neck to the middle of her chest. They would save her life. And then, by chance or fate, the knitting needle would

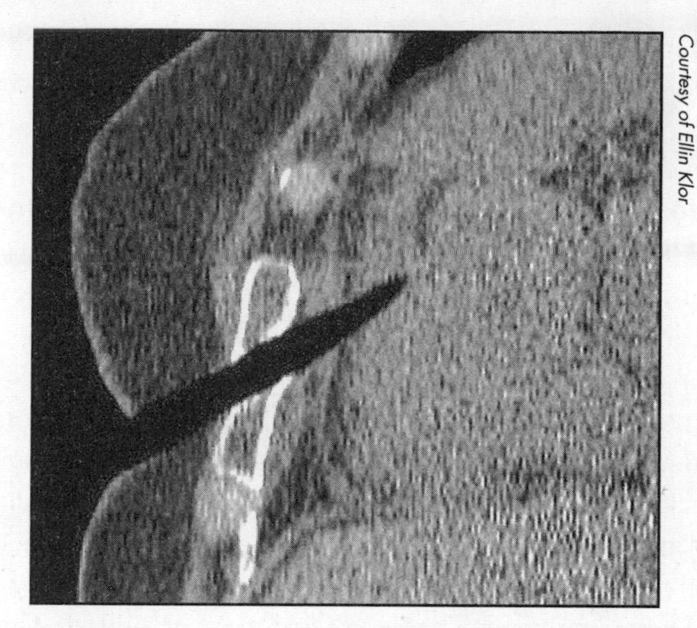

Courtesy of Ellin Klor

CT scan of knitting needle piercing Ellin Klor's sternum and heart.

save her life all over again. In fact, Klor's real struggle for survival was just beginning.

1. The First Rule: Everyone Is a Survivor

On the bright side, it's probably safe to say you're never going to end up with a knitting needle through the heart. But it's equally indisputable that eventually you will face some kind of life-and-death crisis or struggle. Dr. David Spain has a blunter way of putting it. He runs the trauma and critical care department at Stanford Medical Center and sees what happens to regular people all the time. Every day, he says, some of us get dressed, kiss our families good-bye, walk out the door, and get run over by cement trucks. There's no rhyme or reason, but it happens again and again.* I don't mean to depress or scare you. It's just a reality that survivors understand. No matter how hard we dodge, deny, or resist, a cement truck or a hurricane or some other calamity is waiting around the corner for each of us.† Eventually, everyone joins the fellowship of men and women who have been knocked around by life. Admission is inescapable. Membership is inevitable. The first rule of this book is that everyone is destined to become a survivor.

For our purposes, *survivor* is defined as "anyone who faces and overcomes adversity, hardship, illness, or physical or emotional trauma." Survivors keep going despite opposition and setbacks. They may want to quit but they still persevere. Some even manage to excel under the worst circumstances. They make the most of misfortune. They grow in ways they never could have imagined. They don't just exist or subsist. They live fully. In the jargon of the field, they thrive. Whether they survive six months or sixty years, they make the most of their time. *Survivor* comes from the French *survivre,* which means "to live beyond or longer than." It originates from the Latin *supervivere. Super* means "over, beyond" and *vivere* means "to live." Survivors quite literally are *super livers.*

My definition of survivor encompasses people going through difficult times and also the friends and family who stand beside them. In the cancer community, they're called co-survivors or secondary patients. They're the rocks in your life, the ones you grab on to when you're falling down. They're the pals who es-

* Fatal accidents involving trucks are routine, but it's a little surprising how often people run into cement trucks. In May 2007 in Puyallup, Washington, for instance, a forty-seven-year-old man was killed when his Mercedes was struck by a cement truck. In April 2007 in Roseville, California, a sixty-year-old pedestrian was killed by a cement truck. In February 2007 in Washington, DC, a homeless man was struck and killed by a cement truck. And in January 2007 in Saddle River, New Jersey, a retired surgeon driving another Mercedes was crushed by a cement mixer.

† According to *Time* magazine, "91% of Americans live in places at moderate-to-high risk of earthquakes, volcanoes, tornadoes, wildfires, hurricanes, flooding, high-wind damage or terrorism."

cort you to the doctor's office for an MRI. They fix dinner when you don't have energy. They comfort you in the middle of the night when you wake up with paralyzing anxiety. While most attention focuses—understandably—on the person fighting a disease, co-survivors bear a great burden, often silently and without recognition. They suffer a much higher risk of stress, illness, and even death.* In the tiny field of survivor studies, they're pretty much an afterthought, but they know as much as anyone about beating the odds.

If you look around right now—in the coffee shop, airport lounge, or public library—chances are that someone nearby is a survivor. Perhaps the woman next to you is going through chemo and her hair is starting to thin, but she's doing everything she can to look normal. Maybe the man across the way just lost his wife in a car wreck. He's wrestling with depression and wondering how to go on with his life, let alone raise his kids. The guy in the corner may have been let go from his job—he's got no savings—and he doesn't know what to do next. Perhaps the woman across the way is trying to figure out how to help her father with Alzheimer's. Should she put him in a nursing home? How will she afford it?

It's a parallel universe, this unseen world where survivors and co-survivors wage their battles, surrounded by the rest of us, seemingly oblivious. Many survivors describe two coexisting realities. They live with one foot in the regular world and one foot in an invisible realm of hardship and loneliness. Ours may be a confessional culture, but in this other sphere, most people face their struggles quietly, trying not to draw attention. Sure,

---

* Every year, more than fifty million Americans provide care for friends or family with chronic illness or disability. The stress of caregiving can shave ten years off your life, according to one study. Another study of couples over sixty-five shows that if your husband is hospitalized, your risk of dying within thirty days increases 44 percent; if your wife is hospitalized, your risk of death jumps 35 percent. These findings are consistent with the well-known Bereavement Effect: If your spouse passes away, your risk of death within thirty days increases 53 percent for men and 61 percent for women.

some survivors appear on TV, give speeches, and write books, but most don't choose to publicize their ordeals. They endure adversity without talking about it. They don't want to burden anyone else. They don't want pity. They just want it to end.

The desire and drive for normalcy are very powerful. When most people get sick, they want to heal quickly. Knocked down, they try to get back up. For many of us, life is supposed to operate like a seat or tray table on an airplane. On command, it should easily return to its original upright position. Unfortunately, that's not the way it works. The best survivors understand that *normal* is just a fleeting state of mind. Indeed normalcy may seem steady and constant, but it's really just the intermission between the chaos and messiness of life. Survivors accept that life probably won't ever return to the way it used to be. So they let go, adapt, and embrace the "new normal."

Of course, every survivor is unique, and the word itself generates considerable controversy. Some people reject the label *survivor* because they don't want to be branded for life. Like some sort of stigma, they want their ordeal—their cancer, car wreck, or assault—to be expunged. Others prefer to be called *cured* and not reminded of the uncertainty, unpleasantness, and struggle along the way. Some oppose the seriousness and gravity of *survivor*. They dealt with their problem, put it behind them, and press forward without looking back. Others object to the passivity of *survivor* and prefer a more dynamic moniker like *activist, conqueror,* or *warrior.* Some feel that *veteran* best captures their battles and victories, while others believe that *graduate* describes their learning experience and sense of accomplishment. Some abhor what they perceive as the emotionalism of *survivor* with its connotations of heroism, bravery, and courage. In the case of some patients with disease, they prefer the factual phrase *living with* stroke or HIV.

Whichever term you embrace, survival is typically seen as a pass/fail proposition. The medical establishment focused for decades on those who were cured and those who were not. Either you lived or you died. Scientists call this binary thinking. Only two variables matter. *A* = life. *B* = death. You're either one or the other. In reality, of course, survival is messy and complicated—a bumpy road, not a final destination. The path from crisis to normalcy isn't smooth, straight, or one-way. In fact, it's wild and wavering. As the experts say: It isn't linear.

Survivors aren't superheroes who vanquish adversity every time and live happily ever after. If you think they're always triumphant, you're wrong. They're regular people who win some and lose some. They share a mind-set but they don't all possess the same personality. They overcome adversity but they don't necessarily accomplish it the same way. They aren't always adaptable and optimistic; they feel stuck and gloomy, too. They don't always live to a ripe old age; sometimes they only make it a few months. Ultimately, what defines a survivor is the talent for making the most of life, however much remains. Survivors figure out what's right for themselves and their families. They're true to their feelings. They don't necessarily spend every moment fighting, say, Lou Gehrig's disease, or raising money to cure Parkinson's. They have bad days. They struggle. They succumb. But even when they're physically gone, they're still survivors. They remain with us in other, more enduring ways. They're *super livers* even when their time on earth is cut short.

2. The Second Rule: It's *Not* All Relative

It was Friday the thirteenth, and Nando Parrado and his rugby club were flying from Uruguay to Chile for a holiday weekend of sun, fun, and sport. For reasons still unknown, their twin-engine turboprop clipped a craggy peak in the Andes Mountains and crashed onto a glacier. Of the forty-five passengers aboard Flight 571, twelve died on impact and another five perished that first night in the freezing cold at twelve thousand feet. Parrado, a lanky twenty-one-year-old college student, lay for three days unconscious in a coma. His head was cracked open in four places. The other survivors had given up on him, dragging his body to a pile of the dead.

¿Nando, podés oírme? Nando, can you hear me?

Those were the first words Parrado remembers when he opened his eyes on October 16, 1972. Immediately he was puzzled: *Why am I so cold? Why does my head ache so much?* Parrado's hand moved to his temple. He found the ridges of the wounds above his right ear. When he pressed, he could feel what he calls "a spongy sense of give." The sensation was sickening, pressing his shattered skull into the surface of his brain. Parrado soon learned that his mother, Eugenia, had died in the crash. His nineteen-year-old sister, Susy, would later succumb to injuries and bitter cold, slipping away in his arms. Despite the devastating losses, Parrado refused to let himself shed a tear. A voice in his head told him: *Do not cry. Tears waste salt. You will need salt to survive.*

After sixty days on the glacier, facing slow starvation and imminent death, Parrado and two others embarked on a last-ditch "expedition" to save themselves. One of the men turned back after a day, leaving Parrado and Roberto Canessa to climb alone. With only the primitive tools they had scavenged from the shattered fuselage, they scaled a seventeen-thousand-foot mountain and discovered—to their astonishment and disappointment—that they were nowhere near civilization. The plane had crashed right in the middle of the mountains, known as the *cordillera,* a lifeless expanse of ice. But they trekked for ten days with Parrado "pulling like a train, leading all the way," determined to save themselves and the rest of the group back at the crash site. Malnourished and exhausted, they somehow managed to walk forty-five miles through frozen wilderness, guided by what Parrado calls "an indestructible longing for home." When they finally came upon a man on horseback, they knew they were saved, and Parrado scrawled a note for the rescuers:

Vengo de un avión que cayó en los montanos . . .
I come from a plane that fell in the mountains . . .

Later at the hospital, Parrado's father asked: "How did you survive, Nando? So many weeks without food . . ." Confronted with the horror of imminent starvation, Parrado said they had no choice. After intense debate, they decided to harvest and eat the flesh of those who had perished. Without flinching, his father replied: "You did what you had to do. I am happy to have you home."

The sixteen survivors of Flight 572 were hailed as heroes and cheered as celebrities, and today Parrado is a successful businessman and television personality in Uruguay. Married for twenty-nine years with two teenage daughters, he travels the world, racing cars and giving motivational talks. He tells me about one powerful experience in Salt Lake City, Utah. While delivering a speech about his ordeal in the Andes, he noticed a rather unkempt woman crying in the audience. Her hair was untidy; her clothes looked rumpled; and her face was colorless and without makeup. When Parrado was finished, she approached. "You saved my life today," she told him. "I was dead. I was born again today." A few years earlier, she explained, she had accidentally run over her daughter while backing out of the driveway. "I killed my baby," she told him. "I've been dead." She explained that she didn't care about anything anymore. She didn't look after herself. She had stopped living. Parrado didn't know what to say. He pulled the woman into his arms and hugged her with all his might. No, this disheveled woman hadn't survived seventy-two days on a glacier. She hadn't lost half of her family and her two best friends. And yet, he thought,

Could there be any doubt that in the ways that mattered most, she had suffered as much as I had?

Until that moment, Parrado tells me, he had always felt a strange, uncomfortable pride about his survival struggle. Only those who stood on that frozen slab would know the depths of despair, the killer cold, and the horrors of starvation. Adventure magazines had always ranked the ordeal in the Andes at the top of every list of history's greatest survival stories. Movies and documentaries were made about their struggle. With the woman in his arms, however, he discovered something deeper and more universal. "We all, at times, face hopelessness and despair," he writes in his remarkable memoir, *Miracle in the Andes*. "We all experience grief, abandonment, and crushing loss. And all of us, sooner or later, will face the inevitable nearness of death." After hugging the woman for a long while, words came to him and he whispered: "We all have our own Andes in life. You also have your Andes."

When it comes to adversity, it's human nature to make comparisons. Which is worse? Getting trapped in the freezing Andes or accidentally killing your child? These questions are inevitable but lead nowhere. While some challenges appear to be more daunting or excruciating than others, if you're going through your own ordeal, it doesn't make any difference where it ranks on some imaginary Richter scale of survival. The second rule of the Survivors Club is that it's *not* all relative. Sure, adversity comes in many sizes and shapes, but if it's happening in your life—if it's got your undivided attention—if the stakes matter to you—then contrasts are irrelevant. The Big One is happening to you, right here and right now. Relativity doesn't matter. No matter the crisis—on a glacier or in a driveway—the second rule of the Survivors Club means that your challenge is just as big a deal as anyone else's.

3. The Third Rule: You're Stronger Than You Know

The gun at his ear was the first clue, followed by a rough shove into the backseat of the green Mercedes. Terry Anderson remembers thinking: *I am in deep shit. I am in real bad trouble. And it's not going to be over soon.* His instinct was absolutely right. The Associated Press correspondent in Lebanon would be blindfolded, chained to a wall, and held hostage for 2,454 days.

Early on the morning of March 16, 1985, Anderson had just finished playing tennis with a friend in West Beirut. On a narrow road, he encountered three scruffy men with guns. "Get in. I will shoot," one man said, pointing the pistol at his head. He hurled Anderson to the floor and threw an old blanket over him. After a short drive, Anderson was bound in tape, blindfolded with a filthy strip of cloth, and interrogated. Later he was chained to a steel cot with his hands and feet in shackles. He could not stand, let alone sit up straight. He was forced to relieve himself in a putrid plastic bottle next to the bed. After twenty-four days prostrate on the metal frame, Anderson thought he would go mad. He told one of his captors: "I can't do this anymore. I'm not an animal. I am a human being. You can't treat me like this."

"What do you want?" the guard asked.

"A book. A Bible . . . You must loosen these chains. I will go crazy."

The next day, Anderson's restraints were relaxed, and they brought him a brand-new red Bible. They let him take off his blindfold to read for thirty minutes. He savored the smell of the fresh ink, the new binding, and the first words of Genesis: *In the beginning . . .*

When we speak, Anderson is finishing a home-cooked lunch of pasta and salad. He's drinking a glass of South African pinotage, a red wine. He keeps seven hundred bottles in his cellar, and there's room to grow. It can hold three thousand. He lives on a 250-acre ranch in Athens County, Ohio, where life is good.* He boards and trains about a dozen horses. Earlier in the morning, he tried to teach some manners to a two-year-old Missouri fox trotter named Scheherazade. Now he's looking out over a two-acre pond, horse pastures, stables, and paddocks.

I ask him how he and the other hostages survived all those days in captivity. "We all had to reach inside ourselves to find whatever we had," he explains. "It is extraordinary what people are capable of doing." A marine in Vietnam, Anderson was a correspondent on three continents and reported on every kind of natural and human disaster. In his long career in journalism, he regrets that he didn't write more about ordinary people doing extraordinary things. His most fascinating laboratory was his captivity in Lebanon, where he met nine other hostages. He's kept careful track of the others over the years. Of roughly twenty long-term hostages who made it home, Anderson says, one went straight to a mental hospital and never emerged while another spent ten years in and out of institutions. "All of us were damaged in some ways," he says, "but I believe we have recovered well.†

"Survival is one thing," he continues. "Survival with grace and dignity is another." Anderson believes one of the greatest surprises of his ordeal was the way his fellow hostages got through the very worst without compromising their decency and humanity. He remembers some of his worst days when he wanted to give up, when he couldn't face any more abuse, isolation, or the revolting bowls of fatty lamb and rice. "I can't do this, God," he would say. "I'm finished. I surrender."

---

* After his release, Anderson sued Iran for sponsoring his kidnappers. In March 2000, a federal judge awarded him and his family $341 million in damages for their pain and suffering. Anderson says he never received the whole amount but was given a substantial sum from frozen Iranian assets.

† Anderson is quick to point out that at least ten hostages perished or were executed in captivity.

"But at the bottom," he writes in his powerful memoir *Den of Lions,* "in surrender so complete there is no coherent thought, no real pain, no feeling, just exhaustion, just waiting, there is something else. Warmth/light/softness. Acceptance, by me, of me. Rest. After a while, some strength. Enough, for now."

Anderson believes that he reached this state of grace once or twice. "A few hours later, it fades, and the anger and frustration and longing are back," he writes. "But the memory is there, the sense of presence. And sometimes the place is reached again, briefly. Not often, but sometimes.

"Meanwhile, the hours are endured, the days gotten through. And the nights are spent in prayer, and thought, and the effort to get back to that place." We all can find this kind of power in ourselves, Anderson believes. It's there. Inside us. Waiting to be released.

The third rule of the Survivors Club is that you're probably stronger than you know. When you face a real crisis, you'll discover strengths and abilities that you never knew existed. In interviews with survivors around the world, every single one described this phenomenon. Sometimes, they uncovered hidden capabilities they didn't realize they possessed. Occasionally, qualities that they'd always believed were flaws—like stubbornness—ended up saving their lives. The third rule underscores the Japanese proverb that adversity makes a jewel of you. When you're put to the test, you may even be stunned by your own power. It's just waiting to be called into action.

4. The Superhero Next Door

Ellin Klor's home looks like a colorful collage of her many interests: heaping bowls of yarn on the floor, quilts on the walls, framed photos of adventures around the world, stacks of books on almost every surface, and her daughter's eight-harness loom in the living room. Surrounded by this creative clutter, Klor hands me a clear plastic sack that's marked BIOHAZARD SPECIMEN BAG with a bright orange warning symbol. Inside, there's a small test tube labeled #1 FOREIGN BODY. It's a perfect euphemism for the four-inch fragment of knitting needle that lodged in her heart. The splinter is still stained with dried blood. Klor made a special trip to Stanford's pathology department to retrieve it. Some day when she's emotionally up to it, she plans to create a "silly little shrine" with the needle and CT scan of it piercing her heart.

For our lunch date, Klor chooses the Cool Café—named after its chef and owner—on the campus of Stanford University. It's a short drive from her home, and our table looks out on a sculpture garden glinting with Rodin bronzes. Midway into our meal, Klor surprises me by leaning forward to unbutton her linen blouse. I had asked about her scar, and now she's showing me a pink ridge from the base of her throat to the middle of her chest. This is where the trauma surgeons opened her up to remove the knitting needle. Then she points to a lattice of other markings nearby, the telltale tracks of a different and much more serious survival struggle.

It was early Saturday morning, just twelve days after doctors had delicately removed the splinter and stitched her up. Klor had been home for a week, thankful for the attention of her husband and daughter. They had cooked delicious meals, helped change her bandages, and the three were feeling closer than ever. Among her friends and acquaintances, she was a reluctant star. Families from her daughter's school came over with get-well cards and casseroles. She appreciated their concern but was eager to put the ordeal behind her.

Klor awoke that Saturday morning with excruciating chest and back pain. Writhing and struggling to breathe, she had no idea what was happening, and she rushed to the emergency room. Just when she felt life was getting back to some semblance of normal, she was splayed on a gurney again. Doctors poked and prodded her. They listened to her heart and lungs. They whispered their greatest fear: Perhaps it was a pulmonary embolism, a potentially fatal blood clot in her lungs. They ordered immediate scans along with enough morphine to erase the pain. Then the nervous wait for results began. Klor shuddered: *Is this it? Am I going to die?* When the doctors returned, they shook their heads and seemed confused. The tests were all negative. There was no blood clot. Indeed, they couldn't find anything wrong. Her lungs were clear and her heart was healing just fine. So they explained it away as some kind of fleeting discomfort from surgery and gave her more painkillers before sending her home.

The next day, Sunday, she woke up nervous. Nothing hurt at all. She felt so lucky. She could carry on with her life. She could get ready for work. She could launch into a bunch of projects that awaited her attention. But once again, in an instant everything changed. Klor was home alone when the phone rang on Monday afternoon. A radiologist from Stanford wanted to see her right away. At the hospital, the doctors explained the urgency. It had nothing to do with the knitting needle. On a CT scan, the radiologist had detected a mass under her arm. It looked like an enlarged lymph node—a telltale sign of breast cancer. They needed to do a biopsy right away. In the days that followed, Klor was diagnosed with high-grade invasive ductal carcinoma. She had cancer in her right breast.

A decade earlier, she had battled the disease in her other breast, but this was a brand-new cancer. A recurrence of the old tumor would have meant the cancer had spread and that her chances of survival were slim. But this was like starting from square one, a fresh battle. Klor felt so lucky that she let out a whoop when the doctor informed her that only one lymph node was implicated and the cancer could be stopped.

The knitting needle through her heart had actually saved her life, her doctors said. If she hadn't rushed to the ER—if she hadn't been screened with all those machines—the tumor probably wouldn't have been detected until it had grown and spread. Thanks to the needle and the mysterious pain, they had found the malignancy in time and they were confident she would beat it again. *I didn't die from the knitting needle,* she remembers thinking, *so I'm not going to die from cancer.*

Klor spent most of the year undergoing surgery, chemo, and radiation. On every single trip to the doctor, she was accompanied by family or friends. During that time, she also managed to finish a quilt, knit shrugs, scarves, and shawls, and watch her daughter grow up very fast. Klor suffered plenty from the treatments, but she also discovered something she didn't know about herself. She had always struggled with a sensitive nature and tended to absorb a lot of the negative energy around her. At times, she had been vulnerable to depression. Physically, she wasn't very tough, either. "I really have surprised myself," she says about her experience, adding, "I didn't think I had this kind of strength."

Ellin Klor's story captures all three rules of this book. Ultimately, everyone joins the Survivors Club, sometimes more than once. When you're in the trenches, your immediate challenge is the only one that matters. And perhaps most surprising of all, when you're really tested, you'll discover strength where you least expected. Given her fondness for comfort food, Klor admits that she doesn't exactly look like Wonder Woman, but that's the point. The superheroes of survival—the *super livers*—are all around you. They don't call attention to themselves, but they reside next door. They work beside you. In many ways, they're just like you. The rest of this book is devoted to how they do it and how you can, too.

Survival Secrets

Can a 145-Pound Grandmother Really Lift a 3,450-Pound Car?

In November 1988, a Soviet weight lifter named Leonid Taranenko performed an extraordinary feat. In Canberra, Australia, the Olympic champion and super-heavyweight hoisted 586.4 pounds over his head in the clean-and-jerk competition. It was the greatest lift of all time. At age thirty-two, Taranenko weighed 313 pounds, stood almost five foot nine, and trained intensively six days a week. Although his world record was set aside later because of rule changes, no one has ever been able to lift that much again in official competition. In the rarefied and muscle-bound world of weight lifting, some believe that Taranenko is the strongest man ever.

When world records are celebrated, no one ever mentions people like Angela Cavallo, a 145-pound grandmother who never lifted weights or worked out at the gym. On Good Friday, 1982, Angela's son Tony was tinkering underneath a white Chevy Impala. The car was up on cinder blocks and a jack. One of the rear tires was off. Fiddling with the suspension, Tony somehow rocked the 1964 Chevy off its supports. It came crashing down, and he was struck unconscious in the wheel well. Cavallo could see her son's legs sticking out from underneath. So she reached under the car and grabbed hold of the shiny metal fender. The Impala weighed 3,450 pounds, but Cavallo managed to lift it a few inches to "take the pressure off" her son. She nudged him with her foot, but he wasn't moving. "Tony, get out! Get out!" she kept calling, but there was no response.

Cavallo still lives in the same house in Lawrenceville, Georgia, about thirty miles northeast of Atlanta. The memory of her ordeal is just a blur. She doesn't remember how long she stood there with her palms digging into the fender and her muscles tensed. She was fifty-one years old, five foot eight, and didn't exercise much except for a regular swim in the family pool. The grandmother of two girls, Cavallo didn't even carry the groceries at the store. Her husband and son always insisted on doing the lifting. "Adrenaline went through my body," she says. "And you just do it."

Her son emerged unscathed from the accident, and Cavallo is living proof of the third rule of this book—you're stronger than you know. But how much difference does adrenaline really make? In 1960, two scientists in Chicago analyzed the forearm muscles of people experiencing significant psychological pressure. They discovered that flexor power could increase by as much as 26.5 to 31 percent with certain stimulants like adrenaline and amphetamines. The scientists concluded that people usually don't maximize their power because of various "acquired inhibitions." In other words, we're more powerful than we realize but we're rarely—if ever—required to summon that force. As Angela Cavallo says, "When you're a mother, it's born in you."

2

The Statues in the Storm

Why So Many People Die When They Shouldn't

The optimum core body temperature for humans is 98.6 degrees Fahrenheit. If it drops a few digits, everything begins to go haywire. At ninety-one degrees, for instance, you start to hallucinate. At ninety degrees, your heart beats out of rhythm; at eighty-eight degrees, you don't even shiver anymore; by eighty-six degrees, you lose consciousness. Death knocks at eighty degrees.

On the hypothermia scale, Paul Barney came just two degrees from certain death. When rescuers winched him from a life raft in the Baltic Sea, his core temperature measured just eighty-two degrees, four below the usual point of blacking out. And yet the thirty-five-year-old was alert enough to smile and proclaim, "I'm lucky to be alive!" Barney survived the sinking of the *Estonia*, a massive car ferry that went down in a Force 9 gale, killing 852 passengers. Lounging on the couch in his home in Berkshire, England, today, he tells me there are many reasons he lived. Among them, he's thrifty and claustrophobic.

On September 28, 1994, Barney was making an overnight trip from Estonia to Sweden. To save around forty dollars, he decided against taking a cabin belowdecks and instead camped out in one of the open spaces of the sprawling fifteen-thousand-ton ship. Around 11 PM, he settled down in the perfect spot in Café Neptune, the cafeteria on Deck 5 at the stern or back end of the vessel. Outside, the wind and waves were whipping up "rough"—the *Estonia* was pushing through a storm—and Barney was relieved that he didn't feel seasick. At least not yet. He spread his sleeping bag on a bench and nodded off. Around 1 AM, he awoke to a sudden bang. The ship was tilting to the right at an odd angle. Tables and chairs began to slide. At first, he wondered if the ferry had run aground, hitting a rock or island in the Finnish Archipelago. Then he realized the ship wasn't swaying at all—instead, it was listing to one side, and the angle was increasing. "So I thought, *We've got to do something about that,*" he says.

A landscape architect with two master's degrees, Barney was always pretty good at what he calls "orienting" himself. He decided the best place to get more solid footing and figure out an escape plan was the doorway between the café and the exterior promenade deck. As the ship tipped over even more, he maneuvered around the door frame to stay standing. From this precarious perch, he attempted to redraw a mental map of the ship, but his brain struggled to keep up with so much confusing information. With dishes and glasses crashing everywhere, he knew one thing for sure: The boat wasn't suddenly going to right itself. Barney was getting more and more frustrated. There were no emergency instructions from the captain or crew. There wasn't any lifesaving equipment around. "I realized that this was quite a desperate situation," he says, "and I was quite likely to die." Barney expected to see passengers scrambling for their lives. He imagined scenes of bedlam with people clawing for life preservers and fighting for the lifeboats. Instead, he encountered something truly strange: So many fellow passengers seemed unable to do anything at all. "People were just not moving," he says. "They were frozen to the spot, almost waiting to be told what to do." As the lights flickered on and off, they looked like marble statues, pale and immovable.

"Why don't they do something?" he asked an Estonian man who was sharing the door frame with him.

"Just don't think about it," the man replied.

For a minute or two, Barney and the other passenger debated how long they would be able to survive in the Baltic with the water temperature around forty-six degrees Fahrenheit. The Estonian answered: four minutes. From their position, only one life belt was visible. "We both were eyeing it up," Barney says, but it was stuck to a wall across a ravine where the promenade deck used to be. Barney hesitated—it was too far away—while the Estonian made his move. He lunged across the gap, managed to grab hold of the ring, but a wave washed over the stern and swept him away with the life belt.

In that instant, Barney recognized that the sea would soon come for him, too. "I was on my own," he says. So he tried to force every distraction from his mind—the strange groans of the ship, the careening objects—and concentrate on making a plan. He managed to pull some warm clothes from his backpack, and then he focused on finding an escape route. He took off his boots—he didn't want to end up with them in the water—then studied the landscape. The ship was on its side, and a half-moon cast some light on the surreal surroundings. What had once been the ceiling of the promenade deck was now a vertical ladder of pipes and vents. Barney knew this was the way out. He scampered up the trellis and soon found himself all alone on the massive overturned hull of the *Estonia*. Gale-force winds screeched in his face, and waves crashed from every direction, threatening to wash him away, but he remembers no fear. Instead, he was filled with awe. "I'm elated," he says, remembering the moment. "Suddenly I'm free; I'm not claustrophobic; I'm out of there. I haven't escaped but I'm immensely relieved and very happy. It's fairly short-lived but I had time to reflect." He pauses. "Of all the places in the whole world, what a place to be standing in your socks! The storm all around me and the ship nearly disappearing into the sea. It's like a giant surfboard."

At the other end of the ferry, some five hundred feet away, he could see other passengers trying to inflate a life raft. Buffeted by wind and waves, he edged toward them, but the hull was pockmarked with portholes. In the maelstrom, it was impossible to tell which windows were open and which were closed. A misstep could plunge him back into the drowning ship. With great care, he finally made it to the bow.

Barney helped the others launch the life raft, and he jumped in at the last possible moment, but a wave flipped it upside down. It was pitch black underneath, and Barney swam for his life. The sea was alive with passengers thrashing and trying to climb to safety. After righting the raft, he managed to pull himself up and drag some others aboard. All told, there were sixteen soaking, shivering people huddled together now. A lull in the storm calmed the waters, and the seascape looked eerily like the set of a disaster movie with rafts bobbing and beacons flashing. In the red smoky haze, just thirty minutes after beginning to list, the *Estonia* slipped beneath the surface. Of the 989 people aboard, nearly two-thirds didn't even make it off the ship. Many were trapped in cabins or corridors and never had a chance. Others froze like statues. Barney and those who managed to escape now faced thirty-foot seas, sixty-mile-per-hour winds, and a voyage of the damned.

1. The Professor of Survival

Paul Barney's story reflects the blunt reality of survival: Too many people perish when they shouldn't. They morph into marble instead of taking action. Exploring this phenomenon is the main focus of Dr. John Leach, one of the world's leading experts on survival psychology. He has lived in Britain's Lake District for more than twenty years, a paradise of fells, waterfalls, and forests near the Scottish border, and teaches an advanced course in survival psychology at Lancaster University. Pouring himself a cup of tea, he remembers his own unpleasant introduction to the Survivors Club. Leach was changing trains one night in London at the King's Cross Underground station, a sprawling hub that throbs with more than thirty thousand passengers during rush hour. He moved through the packed tunnels in a suspended state of consciousness, the kind we all reserve for the drudgery of commuting. Then he noticed the "thickest, greasiest, most cloying smoke I've ever seen." At first, it didn't make sense. There were no flames—just acrid smoke, like the kind that belches from a ship's funnel. He vividly recalls black streams pouring from the wooden slats of a newsstand. Something was very wrong. Almost without thinking, he found his way up to ground level and hurried to the exit. At the top of the station, near the main ticket hall, he was amazed to see that the great Victorian glass dome had vanished behind an impenetrable ceiling of swirling smoke.

In the cool evening air outside, fire trucks screamed onto the scene. Men and women spilled out in the night, covered with soot and ash. Leach decided to go back inside to help. He had served in the military and was trained in rescue operations. But a policeman stopped him and turned him away. Leach explained his expertise—he wanted to assist—but the bobby wouldn't budge. Stepping back, Leach began to observe the disaster in real time. The date was November 18, 1987.

Today, more than twenty-one years later, most of the memories have faded, but Leach recalls certain images in exact detail. He can still smell the foul smoke, more vile than anything he had ever encountered, and he can hear the wail of a uniformed railway worker: "There are people dying down there." For some inexplicable reason, as the fire spread, trains kept on arriving in the station, disgorging passengers onto platforms choked with smoke. Meanwhile, aboveground, officials unwittingly directed passengers onto escalators that carried them straight into the flames. From his vantage point, Leach recalls two other impressions. First, many commuters followed their routines despite the smoke and fire. They marched right into the disaster, almost oblivious to the crush of people—some actually in flames—who were trying to escape. One woman even approached a manager and asked matter-of-factly: "Does this mean my train has been canceled?" Second, many of the Underground authorities were simply overwhelmed. One official was frozen in fear, unable to do a thing, with tears streaming down his face.

Thirty-one people perished in the King's Cross fire, the worst in Underground history. Investigators would later determine that a lit match had fallen through the wooden escalator treads, igniting grease and trash, spreading to the ceiling paint, funneling flames and smoke upward in what they called a flashover, consuming the crowded ticket concourse above. The temperature in the inferno exceeded sixteen hundred degrees Fahrenheit. It took 150 firefighters to extinguish the blaze, and one died in the rescue efforts. Incredibly, the Underground staff never sprayed a single fire extinguisher or spilled a drop of water on the fire. An official investigation found that they were "woefully ill-equipped" and their overall response was "uncoordinated, haphazard and untrained."

Leach has a name for this syndrome. It's called the Incredulity Response. People simply don't believe what they're seeing. They tell themselves there can't really be a fire in London's busiest tube station. *This really isn't happening.* So they go about their business, engaging in what's known as the normalcy bias. They act like everything is okay and underestimate the seriousness of danger. Some experts call this analysis paralysis. The stress of a crisis can cause the key part of the brain that processes new information to misfire. People lose their ability to make decisions. They turn into statues. The disaster at King's Cross proved to be a seminal experience in the education of John Leach. Indeed, the calamity helped shape one of his central conclusions about survival. He would later write in his blunt prose: "Denial and inactivity prepare people well for the roles of victim and corpse."

2. The Mystery of the Unopened Parachutes

To solve the mystery of who lives and who dies in a crisis, you need to be able to examine people under extreme pressure. But therein lies a vexing problem: The shock of an emergency is almost impossible to capture or measure in real time. Fortunately, psychologists are very creative in their search for answers and go looking in the most unlikely places—like East Troy, Wisconsin, where folks at the tiny municipal airport heard a terrible thud in June 2002. It didn't take long to discover the reason. Near a hangar they found the crumpled body of Luca Bertetto, a thirty-one-year-old engineer from Italy who had recently moved to the area. With thirty-three previous parachute jumps under his belt, Bertetto was last seen at an altitude of three thousand feet plummeting toward earth. He was skydiving along with six other jumpers from the local Sky Knights Parachute Club. No one saw him "go in"—the sport's euphemism for hitting the ground. Investigators found that the handles on his main parachute, emergency cutaway, and reserve chute were in place and had not been pulled. For some reason, the safety device designed to trigger the chute automatically at low altitude had not fired. A coroner concluded that Bertetto showed no signs of medical problems during the jump and died of massive internal injuries. Why did this young man fall from the sky without opening his main or backup chutes? Did he simply forget to pull the handles? Suicide was ruled out as a possibility. The US Parachute Association calls it a tragic mishap. Survival experts believe it's a case study of why people die when they shouldn't and how sometimes we can fail to save ourselves.

James Griffith is one of the country's top experts on what goes wrong when people die skydiving. When we speak, he's just returned from a busy day at the local drop zone in south-central Pennsylvania where he jumped four times from fourteen thousand feet. At age forty, he's a veteran skydiver and part-time instructor with more than three thousand jumps over the past ten years. In his real job as a psychology professor at Shippensburg University, he has studied all of the reports of fatal skydiving incidents going back to 1993. "Every time you jump, you literally are saving your own life," he says. "Each time, you are cheating death in a way." Yes, with good training and equipment, the sport is reasonably safe, "but there's always an element that something could go wrong."

If you examine all of the accidents, Griffith says, it turns out what happened to Luca Bertetto isn't too surprising. There's even a name for it. It's called a no-pull—when the skydiver simply fails to deploy the main or reserve chutes. Another variation is known as a low-pull, when a jumper activates the parachute at a low altitude, often too late for survival. Every year, according to Griffith, around thirty-five people die in skydiving accidents out of some two and a half million jumps. That's one fatality for every seventy-one thousand leaps.* Ten percent of all parachuting deaths—a small fraction—involve no-pulls or low-pulls. So what goes wrong? In short, human error.

After you rule out suicides and physical problems like

* For comparison your chances of dying this year from a regular fall right here on earth—say, down the stairs—is one in twenty thousand.

heart attacks and bumps on the head, in 75 percent of no-pull and low-pull cases, skydivers lose situational awareness. They don't realize their altitude because they're distracted by other things. Most people have very limited "attentional resources," Griffith explains. That means they can only concentrate on a few tasks at a time. If they're busy, say, practicing a new flying technique, they may simply forget to pull the handles. It seems hard to believe, but Griffith says skydivers get so preoccupied with one activity that they fail to deploy their chutes. In addition, "humans are absolutely horrible at telling time." Even when skydivers know to pull their main chutes forty-five to seventy-five seconds after jumping from a plane, they're often way off judging the passage of time. When they finally take action, it can be too late.

Another, more disturbing reason for no-pulls is what skydivers call brainlock. Jumping out of a plane with your heart pounding and stress hormones pumping, it's no surprise that your mind can freeze up for a few seconds. You can literally forget where you are and what you're doing. It happens to all of us every day—our brains seize up—but we're usually sitting at our desks or pushing a cart through the grocery store. When you're speeding 120 miles per hour toward earth, it can be fatal if you don't recover in time. Friends and other skydivers suspect that Bertetto brainlocked on his last jump. It can happen to any parachutist, although how quickly you recover is believed to be a function of how many times you've jumped before.

What exactly is brainlock? John Leach has actually tried to measure it. Along with his colleague Rebecca Griffith (no relation to James Griffith), he tested the memories of forty parachutists at three different stages: right before a jump, after a landing, and on a nonskydiving day. Under stress, he found that people often display memory problems. They seem to forget what they're supposed to do. On the surface, it appears their ability to remember gets overwhelmed by other thoughts, anxiety, and worry. But when Leach probed, he discovered their memories aren't actually impaired at all. They know exactly how to deploy their main and reserve chutes. So what happens? Leach theorizes that their knowledge—how to save themselves—is stored in their long-term memory, but under great stress, that information can't get across to the part of the brain where it's activated and put to use. Leach found that this happens to novice and experienced parachutists alike.

So what lessons can you draw from the mystery of the unopened parachutes? What can you learn for your own survival here on the ground? Christian Hart is a psychology professor at Texas Woman's University who works with James Griffith trying to analyze what goes wrong in the so-called red zone below twenty-five hundred feet on the altimeter. A veteran of more than four hundred jumps, Hart has interviewed skydivers who didn't pull their chutes and were saved just seconds before impact by their automatic activation devices. He has also reviewed many reports that skydivers have filed about these harrowing incidents. He believes two kinds of personalities emerge under extreme pressure. The first type keeps trying to solve problems no matter what happens. They refuse to quit and sometimes die trying to save themselves. The second type gives up quickly. They resign themselves and surrender.

Griffith and Hart believe that parachuting offers three survival lessons for the rest of us who don't jump out of airplanes. First, try to relax. Some skydiving instructors have a special signal when they're free falling with anxious students: They pat the top of their heads. It's a sign to stay calm. The simple act of remembering to loosen up can break you out of brainlock. Second, remember where you are. It may seem obvious, but situational awareness can mean the difference between life and death, whether you're hurtling toward earth at terminal velocity or driving seventy-five miles an hour on the interstate. Third, never give up. Many parachuting deaths could have been prevented if skydivers kept working on their problems. Human and mechanical errors are fixable, but you never find out if you give up.

3. Mr. Positive and His Fatal Flaw

Let's return now to that lifeboat in the stormy Baltic where Paul Barney was surrounded by a "sickening soup" of dying men and women. Cold, wet, and miserable, Barney clearly remembers one passenger whom he nicknamed Mr. Positive. As waves pounded the craft and froze its passengers, Mr. Positive blurted out all sorts of cheery thoughts. He was "quite a vociferous character," Barney recalls. "'We're going to be saved,' he would say. 'They're coming for us. It won't be long.'" Sometime before dawn, Mr. Positive fell silent, succumbing to the cold. "Sadly, his positivity ran out," Barney says.

"I think he overcooked it," Barney goes on. "Emotionally, he let himself release too much. He was let down too often. And that would have taken its toll on him quite a lot. Obviously you're on an emotional roller coaster—big time—in that situation, and he would have been . . . raising his hopes and having them dashed time and time again. And that would have really stripped him of his energy." This phenomenon is known as the Stockdale Paradox, named after Admiral James Stockdale, the highest-ranking American prisoner of war in Vietnam. The idea was popularized by author Jim Collins in his best-selling book *Good to Great*. When Collins asked Stockdale to explain which American prisoners perished in captivity in Vietnam, the admiral replied, "Oh, that's easy. The optimists." Collins was perplexed, but Stockdale explained that the optimists "were the ones who said 'we're going to be out by Christmas.' And

Christmas would come, and Christmas would go. And then Thanksgiving, and then it would be Christmas again. And they died of a broken heart." Stockdale went on: "This is a very important lesson. You must never confuse faith that you will prevail in the end—which you can never afford to lose—with discipline to confront the most brutal facts of your current reality, whatever they might be."

Paul Barney instinctively understood the Stockdale Paradox: He always faced stark reality but never gave up hope. I ask him if he can describe Mr. Positive's face and features. "No, I don't remember anyone's appearance from the life raft journey," he says. "You don't actually remember faces. You don't take on board faces. If you take on board a face, somehow it seems like you make some sort of emotional connection and that can be too dangerous." He pauses for a moment. "If you make an emotional connection with a face and they're lost, whatever, then you're likely to go as well. I remember faces before and I remember faces afterward, but during the whole event, you just don't take anything like that on board."

Barney carefully managed his own emotions that night, but he also tried to comfort his terrified boat mates. In particular, he attempted to help an Estonian woman in her early twenties who was totally hysterical. He huddled with her for a long time, stroking her dark hair, trying to calm her down and keep her warm. But a wave flipped the life raft and tossed them into the sea. When they managed to crawl back, the woman ended up on a different side, and Barney focused his attention on tying himself to the raft because he didn't want to go overboard again. During the night, the young woman died from the cold.

Wave after wave swept over the raft, washing away any warmth the survivors could summon. Some passengers shook violently from the cold—screaming and thrashing until they lost consciousness. Others stared straight ahead like zombies, unresponsive, deadened to the agony around them. Two crew members from the *Estonia* were no help at all. Even though they were protected by waterproof survival suits, they were in total shock. "They were really not functioning," he says. "They were completely statuesque."

Perhaps it was just a hallucination or illusion brought on by hypothermia, but Barney believes the line separating life and death in that boat was very small. Terrified and freezing, he says, "Death seemed like a warm, safe place." He continues: "The overwhelming desire is to put your head down on the nice soft side of the life raft and fall asleep. And that's instant death." So he refused to rest even for a moment. "I was very calculating. I had to keep myself alert and awake at least, and I was always looking out for the next thing that was going to save my life. So that was the idea. It was always one next thing."

Gasping for breath with his mind muddled by shock and cold, he still tried to solve each of his problems. Relying on previous experiences in difficult situations and extensive yoga training, he concentrated on slowing his breathing and reducing his heart rate. The results were immediate. "It helped clear my head," he says, "and I could start thinking and functioning again." Above all, he was driven by a powerful sense of purpose. "I just felt I hadn't achieved everything I wanted to do in life, and there was no way I was going to fall asleep and die of hypothermia in the middle of the Baltic." He also thinks he had an edge that night because of his adventures around the world as a solo traveler. Those trips forced him to become more self-reliant and make quick, intelligent decisions. As the ship sank, Barney understood what he needed to do and took decisive action. In the freezing life raft, he was ready because he was familiar with shock and hypothermia. On a trip to Wales, his hands had been crushed by a sash window that fell on him like a guillotine. Trekking in Norway, he had suffered bitter cold and frostbite. Obviously, those experiences paled compared with this horror in the Baltic, but his life training—as the military calls it—gave him a small advantage.

When dawn finally came, a rescue helicopter swooped from the sky. Barney refused to believe the ordeal was over, and sure enough, as the chopper winched one survivor to safety, the man panicked and dropped from the cable into the stormy seas. For some inexplicable reason, the helicopter flew away, and the man drowned. "That wasn't very good for morale," Barney recalls. Within fifteen minutes, another copter arrived. Barney was the last man pulled to safety. Understandably, after six miserable hours in the lifeboat, he didn't let down his defenses until he was inside the helicopter.

"There's a point where you hand over responsibility," he says. It's like passing the baton, and for the first time, you let your guard down. "It's a dangerous moment," he explains. Indeed, as soon as he began to relax, his body reacted violently, seizing up in terrible cramps. "I was in agony for an hour in the helicopter."

While other ferries had turned around or reduced their speed that night in the Baltic, investigators found that the *Estonia* had sailed right into the storm. Despite controversy over what really happened, authorities believe the ship's gigantic front cargo door broke off under the assault of so many waves. With its mouth open to the sea, the *Estonia* simply drowned. Seawater poured into the car deck, some twenty tons per second, throwing the ship off balance. Only 137 out of 989 people on board survived the disaster, and of the original 16 in Barney's life raft, just 6 were saved. Helicopter rescue pilots saw as many as forty life rafts on the water, but most were empty.

4. The Theory of 10–80–10

Lancaster, England, an old city on the Lune River, may not seem like a center of the world, but it's arguably the capital of survivor studies. This is where John Leach teaches and writes papers cited in almost every important study of survival. Across the world in Spokane, Washington, home of the US Air Force's Survival School, experts memorize and rattle off Leach's most important findings. They revere him like some kind of prophet and treat his book, *Survival Psychology,* like the bible of who lives and who dies.

Leach isn't just an academic. He has a second life that is never mentioned in his biography or his work. Leach served as an officer in the Royal Air Force, remaining in the reserves, and was mobilized to Iraq in 2003. He specializes in teaching British forces how to survive in every environment. His own training has taken him to the freezing reaches of Canada, the steamy jungles of Belize, and the scorching deserts of Southern California. When he isn't working with British forces, he spends his time observing and analyzing people in crisis. Life is his laboratory. For instance, on a recent flight from Blackpool to London, he sat down and checked under his seat for the life vest. To his surprise, it wasn't there. He summoned a flight attendant, who refused to believe the life preserver was missing. She looked herself, admitted it was gone, and went searching for another. The woman sitting next to Leach watched the entire exchange, but when he suggested that she check under her seat, "she resolutely refused." Leach says his fellow traveler was "going out of her way *not* to consider the possibilities." Perhaps she didn't realize this flight would take her over the chilly waters of the Irish Sea, where a life vest could come in handy. Perhaps she believed that everyone dies in airplane accidents so it wouldn't make any difference. Or perhaps she didn't want to be bothered bending down and touching the grimy underside of a budget airline chair. For whatever reason, the woman was making a potentially deadly mistake. The illogic goes like this: If you don't think about something, it won't happen. If you do think about it, it will. So when it comes to danger, it's better not to think at all. It's classic superstition, Leach says, and nothing could be more perilous to your survival.

In any emergency, he continues, people divide into three categories. First, there are survivors like Paul Barney who manage to save themselves. Second, there are fatalities like the passengers on the *Estonia* who never had a chance and died immediately when the waters came. Third, there are victims who should have lived but perished unnecessarily. In many circumstances, Leach says, there is no way to stay alive. He calls this the Reality Principle. "There are times when you have no choice," he says. "You die. Full stop. It's just the way it goes." You may do everything right to save your life, but some crises simply aren't survivable. They're fatal from the outset. Thus, a certain number of people in the King's Cross fire were doomed from the moment the match fell through the escalator. They didn't do anything wrong. They lost the "cosmic coin toss." It just wasn't their day.

Leach feels great sympathy for these fatalities, but they're not especially interesting from a psychological or survival perspective. Their actions—their frame of mind—make no difference in the outcome. Blame fate or chance, but they were never going to live. From a scientific perspective, that leaves two groups to examine: the survivors and the victims. For years, Leach was preoccupied with the survivors and what makes certain people like Paul Barney excel in extreme situations. Why do they perform so effectively and decisively when others freeze? How do they control their emotions and organize their plans?

Leach and his team methodically interviewed Barney and the survivors of many disasters. He tested them for key psychological traits. His goal—call it the Holy Grail of survival—was to identify what makes some people *super livers*. Leach wasn't making much progress until March 1987 when another European car ferry, the *Herald of Free Enterprise*, capsized just outside the Belgian port of Zeebrugge. Watching news reports of the disaster, Leach realized he was asking the wrong questions and focusing on the wrong group. The *Herald of Free Enterprise* sank less than a mile from its dock. The waters were shallow, and rescue operations began almost immediately. And yet, out of 539 people on board, 193 died. An official inquest would find that the front doors of the *Herald of Free Enterprise* had not been closed before setting sail because of negligence and "a disease of sloppiness." Blame ultimately fell on the owners of the ship, its captain, its first officer, and a crew member named Mark Stanley who was supposed to have shut the doors but instead was napping in his cabin when the ferry left port.

At his home in the town of Sedbergh, Leach was baffled. How could so many people drown so close to shore? A rescue helicopter had arrived in less than half an hour. The water wasn't very deep. The weather wasn't dangerous. So what happened to all those people who didn't make it? "That's when I realized I had been asking the wrong question," he remembers. "The question I had been asking was, *How can a few people survive these extraordinary circumstances?* The question I should have been asking is, *Why do so many people perish when they don't need to?*"

Instead of focusing on survivors, Leach now focused on victims, the ones who panic, freeze, or surrender. His goal was to understand how some people actually "become prone to dying." There's "nothing magical or metaphysical" about this process, he says. "It's all about the engineering system called your brain. In the simplest terms, something becomes impaired—dysfunctional—and that means you don't make the right decisions."

After examining countless disasters and categorizing the ways people respond to life-threatening situations, Leach came up with what might be called the Theory of 10–80–10. First, around 10 percent of us will handle a crisis in a relatively calm and rational state of mind. The top 10 percent are the survivors like Paul Barney. Under duress, they pull themselves together quickly. They assess situations clearly. Their decision making is sharp and focused. They're able to develop priorities, make plans, and take appropriate action. As the *Estonia* began to list, Barney instantly analyzed his predicament and devised a survival strategy. He also kept his emotions in check so that he could think and solve problems. He refused to let himself get overwhelmed. Psychologists call this process "splitting," and it's common among people who keep their cool under the greatest stress.

If all this sounds hard to do, you're right. Most people don't react as purposefully as Barney in the chaos of the *Estonia*. Leach says the vast majority of us—around 80 percent—fall into the second band. In a crisis, most of us will "quite simply be stunned and bewildered." We'll find that our "reasoning is significantly impaired and that thinking is difficult." We'll behave in "a reflexive, almost automatic or mechanical manner." Leach points to the commuters in the King's Cross who trudged forward with their routines despite fire and smoke. Under tremendous pressure, most of us will feel lethargic and numb. We'll sweat. We'll feel sick. Our hearts may race. And we'll experience "perceptual narrowing" or so-called tunnel vision. We'll stare straight ahead. We'll barely hear people around us. We'll lose sense and sight of what's going on around us. In short, most of us will turn into statues in the first moments of a crisis. It's okay—it's not necessarily fatal—and it doesn't last forever. The key is to recover quickly from brainlock or analysis paralysis, shake off the shock, and figure out what to do.

The last group—the final 10 percent—is the one you definitely want to avoid in an emergency. Simply put, the members of the third band do the wrong thing. They behave inappropriately and often counterproductively. They make the situation worse. They're the ones who lose control of themselves. In plain terms, they freak out and can't pull themselves together. On Barney's life raft in the Baltic, the hysterics didn't make it through the night.

If you think about your own response in stressful situations, you can try to predict whether you fall into the top 10, the statuesque 80, or the dangerous 10. In reality, however, it's very hard to know what you'll really do. Even highly trained professionals, like policemen or firefighters, sometimes freeze, while children with no training glide through unscathed. In part 2 of this book, you'll have the chance to test your Survivor IQ and find out more about your survivor personality. No matter which category you fall into, experts say, you can definitely practice and improve your crisis response. With some effort, you can master the habits of effective survivors and also learn what *not* do in a crisis. Perhaps the best place to start is by stepping aboard United 232, a DC-10 cruising at an altitude of thirty-seven thousand feet. It's a clear blue summer day, perfect for flying.

Survival Secrets

The Human Popsicle

Dr. Gordon Giesbrecht hates the cold but that hasn't stopped him from dunking himself in ice, injecting freezing water into his veins, and lowering his body temperature to the point of hypothermia. Giesbrecht keeps meticulous track of his self-inflicted torture: He's gone hypothermic thirty-nine times. Occasionally he even takes his experiments outside, braving the frozen lakes of Canada, deliberately skiing or snowmobiling off the ice into the frigid water. Until he came along, the absolute limit for human hypothermia research was ninety-five degrees. But that was far too warm, Giesbrecht believed, plunging his own temperature in one experiment to 88.16 degrees. This masochism—all carefully supervised—has made Giesbrecht one of the world's leading experts on freezing to death. It has also earned him the nickname Professor Popsicle.

The mild-mannered fifty-one-year-old physiologist is director of the Laboratory for Exercise and Environmental Medicine at the University of Manitoba in Winnipeg. He started freezing himself in June 1986 and hasn't really stopped since. His goal is to get up close and personal with hypothermia and to understand why so many people freeze to death every year. In the United States alone, at least seven hundred people die annually from exposure. *Vitas Salvantes* is the Latin motto of his lab. It means "Saving Lives."

Above all, Giesbrecht wants to explode the myths of the cold. First and foremost, he wants to dispatch the belief that hypothermia kills quickly. "If you think you have just minutes to live, that becomes a self-fulfilling prophecy," he explains. "If you decide you're going to die, you tend to panic and then you do things that are more likely to bring about a negative result." Some 95 percent of those who perish in cold water aren't actually hypothermic, he says. In fact, their body temperatures turn out to be almost normal. The cold doesn't kill them. It's the terror that leads to drowning and heart attacks.

So what should you do if you end up in cold water? Giesbrecht recommends a straightforward 1–10–1 system: You have one minute to get your breathing under control, ten minutes of meaningful movement, and one hour before you lose consciousness. "Survive the first minute," he says, and you're on your way to saving your life. The most immediate danger comes from what's called cold shock. This includes a gasp reflex followed by uncontrolled breathing known as hyperventilation. As you gasp for air, you're more likely to inhale freezing water. This response also makes it very difficult to coordinate your swimming movements. Your first goal is to fight the panic and get control of your breathing. Next, you've got ten minutes to move—swim to safety and crawl out of the water. After ten minutes, your muscle and nerve fibers get so cold that they don't function anymore, and you can't move. If you're running out of time and can't climb out, you should try to freeze your arms to the ice so that when you eventually lose consciousness, you won't sink to the depths. Finally, you've got about one hour before you lose consciousness. If you've got some extra body fat, you're in luck. The heavier you are, the more time you've got in the cold. For a slight, lean person, hypothermia can set in after forty minutes in forty-six-degree Fahrenheit water. At 190 pounds and more than six feet tall, Giesbrecht says it takes his body about an hour to become hypothermic at this temperature. Babies, children, and small adults are especially vulnerable because they loose heat faster than bigger people.*

Another myth of hypothermia: Most heat loss comes from your head. That's a falsehood perpetuated by mothers and grandmothers who want their children to wear hats. In fact, Giesbrecht says, only 8 to 10 percent of your heat emanates from your head. That leaves 90 percent from the rest of your body. In cold water up to your neck, much more heat escapes from the rest of your body than your head. Shivering, he adds, is your body's way of turning up

---

* A major determinant of heat loss is the surface area of your body relative to your weight. More surface area means more heat can escape. Relative to their weight, newborns have about three times more surface area than adults and are therefore much more vulnerable to hypothermia.

the thermostat, producing up to five times more heat than simply resting in a warm environment.

A third myth: Hot drinks like cocoa and coffee help you beat the cold. In fact, Giesbrecht says, any drink with a lot of sugar will make a bigger difference. "The heat in drinks has mostly a psychological benefit," he explains. "One or two cups of warm fluid won't harm you very much." But sugar helps a lot, providing fuel for the body to generate heat and fend off the cold.

At his laboratory in Winnipeg, volunteers earn a hundred dollars for a dunk in the ice tank, and after ten dips, they win a coveted place in the exclusive (and facetious) Polar Bare Society. Their photos go on his wall along with all the others who have frozen themselves in the name of science and saving lives. As for his own future, Giesbrecht is thinking about retiring from those frigid plunges. Thirty-nine times is enough and, most of all, Professor Popsicle just hates the cold.

3

Ninety Seconds to Save Your Life

The Wrong (and Right) Things to Do in a Plane Crash

A woman screamed, a child cried, and Jerry Schemmel told himself: *We're going down.* The words *bang* or *boom* don't do justice to the earsplitting explosion aboard United Flight 232 en route from Denver to Chicago. The plane shuddered, and Schemmel imagined the worst. Just seven months earlier, Pan Am 103 had blown up over Lockerbie, Scotland. Now Schemmel feared it was happening again and he was going to die with 295 other passengers and crew.

July 19, 1989. From the start, it just hadn't been his day. Schemmel's first flight had been canceled for mechanical reasons, and he spent the next five enervating hours on standby at Stapleton International Airport. As deputy commissioner of the Continental Basketball Association, the NBA's farm league, he was heading to the annual player draft in Columbus, Ohio. The exasperating morning had drained him, but as soon as United 232 took off, he leaned back in seat 23G and let go of all the stress. The cabin was filled with a summery feeling. Flight attendants handed out a picnic lunch of chicken strips, chips, apples, and cookies.

At 3:16 PM central time, United 232 was making an easy right turn at thirty-seven thousand feet when the huge fan rotor in engine two disintegrated. The metal disk is the size of a full-grown man, and as it broke apart, shrapnel ripped through the empennage or tail of the aircraft, destroying all of its hydraulic systems. Inside the cabin, the destruction sounded like a thunderclap. Schemmel immediately began to pray. His first words to God weren't for deliverance or for himself. Rather, he thanked the Lord that his wife, Diane, wasn't with him on this flight. He would never want her to experience this kind of dread. Then he surveyed the rest of the passengers. "What I saw was fear, even terror in their eyes. There was not a contagious sense of panic, just a plane filled with very scared people."

Within a few moments, Captain Al Haynes came on the intercom. In a calm and commanding voice, he explained there was a problem with one of the plane's three engines. DC-10s were capable of flying with just two engines, he said, but their arrival in Chicago would be delayed. Despite the announcement, Schemmel had a gut feeling the crisis was much worse than anyone realized. A few anxious minutes later, Captain Haynes was back on the intercom. His voice was still steady and reassuring, but the news wasn't good. The explosion had caused extensive damage to the rear of the plane. Worst of all, the captain admitted, he was having trouble flying the aircraft.

A wave of sobbing and crying swept through the cabin. The DC-10 was still in the air—not plunging or cork-screwing—but now everyone readied for the worst. Quickly, methodically, the flight attendants started to prepare for an emergency landing, and once more Schemmel turned to the Lord. Again, he didn't pray for himself. "Help them, God," he said. "Please help the flight attendants." The safety briefings were short and focused. Brace for impact. Identify the nearest exit. Wear shoes during landing. Schemmel forced himself to repeat the safety information, and then his thoughts turned inward. "My life was basically in order, at least as much as anyone has a right to expect when suddenly faced with death," he recalls. *"Take me, God, if You have to,* I prayed calmly and silently. *I'm ready."*

Schemmel was twenty-nine years old. He had grown up in Madison, South Dakota, in a big household with six siblings. His parents had taught him focus, discipline, and to think about other people before himself. He had also been a high school quarterback who liked to draw plays in his mind. Now he focused on what he would do when the plane hit the ground. He memorized the layout of the cabin, made a mental map of his different routes to the exits, and examined the safety card. "I would probably die in the crash, but just in case I didn't, I wanted to be ready," he says. *"Don't panic,* I told myself. *Stay calm. Help other passengers out of the plane. Don't flee the aircraft."* His mental preparations were interrupted by the captain, who said he would give them a final command—"Brace! Brace! Brace!"—thirty seconds before touching down. The pilot added: "And folks, I'm not gonna kid anybody. This is gonna be rough."

To fly a plane, even a two-seater, you need a variety of controls like a rudder to regulate what's known as yaw, the vertical angle of the aircraft. To roll left or right, you need ailerons or spoilers on the wings that move the plane from side to side. To go up and down, changing the so-called pitch, you need elevators that are usually located in the tail. To slow down, you need flaps to increase drag. Now imagine flying a 360,000-pound plane without any controls at all. No rudder for aiming the nose. No ailerons for banking. No elevators for climbing or descending No flaps to decelerate. And if you make it to a runway, no nose wheel for steering or brakes for stopping.

As a Marine Corps aviator and experienced airline captain with over twenty-seven thousand hours in the cockpit, Al Haynes had handled just about every imaginable flight problem. But this situation was unfathomable. The plane's three hydraulic systems were designed to work independently and back one another up in the event of failure. Losing all of them simultaneously was supposed to be impossible. The chances were one in a billion. Without hydraulics, Haynes had no way to control the plane's flight surfaces.

From the moment of the explosion, Haynes tells me that there was no time for fear or panic. He and his crew quickly realized the only way to guide the plane was by adjusting the thrust of the two remaining engines. To turn, he accelerated the power on one wing and then the other, literally skidding across the sky. Trying to land a DC-10 this way, he tells me, is like racing a car down Pikes Peak in the Rockies—altitude 14,115 feet—and steering "by opening and closing the doors." To guide the plane to Sioux City, which was seventy miles away, Haynes made a series of sweeping right turns. Each giant loop brought them closer to the airport, but lining up the jet for landing was almost impossible. When the control tower finally told Haynes, "You're cleared to land on any runway," he couldn't help laughing. "Roger," he said. "You want to be particular and make it a runway, huh?"

Just forty-four agonizing minutes after engine two exploded, United 232 made its final, harrowing approach to the airport. With no flight controls, the plane's rate of descent was six times faster than the usual three hundred feet per minute. When it slammed into Runway 22, it was going around 247 miles per hour, more than 80 above a safe speed. The right landing gear, wingtip, and number three engine smashed into the ground, rupturing the fuel tanks. The impact snapped off the tail; the nose of the plane bounced on the runway three times, and the fuselage began to break apart.

The cockpit and right wing sheared off, and the biggest piece of the plane skidded sideways and then rolled upside down. Instantly everything burst into flames.

1. The Myth of Hopelessness

True or false: When a plane crashes, everyone dies. If you're like most people, you answered true. After all, the biggest disasters are tattooed in our memories. *May 1996:* ValuJet 592 slams into the Florida Everglades with 110 people on board. No survivors. *July 1996:* TWA 800 explodes with 230 on board. No survivors. *September 1998:* Swissair 111 crashes near Nova Scotia with 229 on board. No survivors. *October 1999:* EgyptAir 990 dives into the Atlantic Ocean with 217 on board. No survivors. It's easy to think the chances of surviving a plane crash are hopeless. Indeed, during the safety briefing on a recent flight, a fellow passenger whispered to me: "If this plane goes down, we're all dead and there's nothing we can do about it."

I'm embarrassed to admit that every time I fly, I go through a litany of superstitious rituals. I always tap the right doorjamb of the plane when I step aboard. During takeoff and landing, I mumble a short prayer that I learned long ago in Sunday school. So imagine my surprise—and relief—when I discovered that one of the world's leading authorities on plane crashes is also afraid of flying. Arnold Barnett is a sixty-year-old professor at MIT who specializes in operations research, a field of applied math that uses numbers to improve complex systems like air traffic control and assembly lines. Barnett tells me that his interest in aviation safety grew from his flying phobia. Nearly two decades ago, he figured he might calm his fears by learning more about the risks of dying in a plane crash. He started out by asking: Why do people perceive the danger to be so great? Barnett studied the front page of *The New York Times* and found the answer.

Page-one coverage of airplane accidents was sixty times greater than reporting on HIV/AIDS; fifteen hundred times greater than auto hazards; and six thousand times greater than cancer, the second leading killer in America after heart disease.

Next, Barnett created a brand-new measurement that captures precisely what people want to know: *What are my chances of dying on my next flight?* In the aviation safety field, it's known as Q: death risk per randomly chosen flight. Analyzing all the data from the last ten years, here's Barnett's bottom line: When you get on your next domestic flight, your chance of being killed—your Q—is *one in sixty million*. That means you could fly *every day* for the next 164,000 years before you would perish in a crash. No matter how frequently you travel, your risk of death remains the same: *one in sixty million*.* I ask Barnett if his Q research has cured his fear of flying. Not entirely, he laughs, but it definitely helps.

Even *if* you somehow ended up in a plane crash—a remarkably unlikely *if*—your chances of dying are still unbelievably small. Believe it or not, the survival rate in plane crashes is 95.7 percent. Yes, 95.7. More precisely, the National Transportation Safety Board analyzed all the airplane accidents between 1983 and 2000. Some 53,487 people were involved in those incidents, and 51,207 survived. Hence, the survival rate of 95.7. The safety board judged twenty-six of the accidents to be the worst, meaning that they involved fire, injuries, or substantial damage. Excluding those in which no one had a chance, the survival rate in the most "serious" accidents was 76.6 percent. This means that even in bad crashes, more than three-quarters of the passengers live. "Contrary to public perception," the board concluded, "the most likely outcome of an accident is that most of the occupants survived."

---

* For comparison, Barnett estimates your risk of death per automobile trip is around one in nine million, or nearly seven times greater than the risk per domestic flight.

One dangerous consequence of the Myth of Hopelessness is that when people believe there's nothing they can do to save themselves, they put themselves at even greater risk. Before flying, they pop a few drinks in the bar. As soon as they get on the plane, they take off their shoes, crack open a book, read the paper, or crank up the iPod. They put on their face masks to sleep and ignore the safety briefings and information cards. If the plane crashes, they figure it doesn't matter if they're drunk, barefoot, and blindfolded: They're dead anyway.

Here again, consider the facts. According to the European Transport Safety Council, 40 percent of the fatalities in plane crashes around the world occur in situations that are actually survivable. In other words, out of an average of fifteen hundred total fatalities, some six hundred people die in accidents where they might have lived. The question is: Why? Broadly speaking, the planes are well made—the safety equipment is good—the standards are high. Of course, there are plenty of improvements that would make airplanes even safer (like air bags, three-point safety harnesses, and rear-facing seats). But in survivable crashes, the experts say, it all comes down to human factors and what you do—or don't do—to save yourself. Here again, there's another important myth to debunk, so let's return to United 232, burning in an Iowa cornfield, and a powerful example of what people sometimes do right in emergencies.

2. The Myth of Panic

Jerry Schemmel was hanging upside down, suspended in the air by his seat belt. When the DC-10 first hit the ground, a woman still strapped in her seat had flown right by him. A ball of fire roared down the aisle. Then the plane flipped over and he felt searing pain through his spine and neck, as if he had been electrocuted. Dangling in his chair, Schemmel says, "I simply and calmly wondered, *Dead or alive?*" When flames singed his knuckle and he felt the pain, he knew the answer. Somehow, he released his seat belt and fell onto what a moment ago was the ceiling of the plane. Standing up in the midst of the wreckage, he could see other passengers still belted in their seats hanging from above, arms and legs flailing, and blood streaming down like rain.

All of Schemmel's mental planning and preparation suddenly became irrelevant. "There was nothing left to do but react instinctively to the raw physical chaos," he says. *Forget the dead. Help the others,* he told himself. Groping through smoke and twisted metal, he released the seat belt for an elderly man and aided a woman with broken legs. When he spotted sunlight through an opening in the fuselage, he realized it was a way out. "I knew at that moment I was not going to die," he says.

Two men were assisting people through the hole in the plane, and Schemmel remembers, "There was an amazing sense of calm, of order, to the impromptu evacuation." When he finally stepped out into the brilliant light of the July day, he was stunned by the softness of the earth underfoot. The plane had come to rest in a cornfield, but might now explode. His first instinct was to run from the aircraft. He took a few steps, then he heard a sound that stopped him cold. "It was a human voice," he remembers, "a muffled cry that could only be a baby. And clearly, it was coming from inside the wreckage."

Automatically, Schemmel turned and went right back into the fuselage. The fire and smoke were so intense that he couldn't see a thing, and simply followed the sound of wailing. "Keep crying," he said. "Please keep crying." He tracked the noise until he sensed that he was directly above it. Then he started digging. First, he pulled away a duffel bag. Then, a blanket. His hands found an overhead bin that was now on the floor. He reached inside, discovered a tiny arm, and pulled out a little body. Blinded by smoke, he ran from the plane and across the cornfield. When he finally came to a stop, he saw it was a baby. She was wearing a blue dress. She smiled at him. He would learn later that her name was Sabrina Michaelson. She was eleven months old. Remarkably, the rest of her family had survived, too, but—unable to find her in the debris—they had been forced to flee the fire and smoke. Of 296 people on board United 232, there were 185 survivors and 111 fatalities, and investigators determined that passengers lived and died right next to each other in adjacent seats.

The word *panic* derives from the Greek god Pan, who ruled the woods and fields and stirred up mysterious sounds causing "contagious, groundless fear in herds and crowds or in people in lonely spots." As we have seen, panic is the archenemy of survival. Sudden, unreasoning, and hysterical fear that spreads quickly can't be a good thing in plane crashes or ferry disasters. Fortunately, it almost never happens. In fact, researchers examining crises as disparate as the London Blitz in World War II, the Kobe earthquake in Japan in 1995, and the attacks of September 11 have discovered quite the opposite. People rarely lose total control and run around mindlessly. Rather, most freeze until they're told what to do. And some actually manage to keep their wits and act purposefully, even engaging in what's known as situational altruism, like Jerry Schemmel venturing back into the burning wreckage to rescue Sabrina. One semantic distinction is important here. In a burning plane, running for your life, clambering over seats, and fighting for an exit may seem like panic, but it isn't. It's an entirely rational fear-driven response to being trapped inside an aluminum cylinder that's filling with smoke. It's known as panic *flight*.

Lee Clarke, a sociologist at Rutgers University, has examined many potentially panic-inducing situations. It's a widespread misconception, he says, that in fires and other conflagrations, the rules of society crumble and "our true animalistic" selves come forth. "I'm no Pollyanna," he goes on. That kind of "self-interested bastard" behavior definitely exists in the world. But in times of hazard, he argues, "this other human nature kicks in." Social values like altruism are heightened in emergencies, not lessened. "After five decades of studying scores of disasters such as flood, earthquakes and tornadoes," he writes, "one of the strongest findings is that people rarely lose control." In fact, "the more consistent pattern is that people bind together in the aftermath of disasters."

So if people don't often panic in the crazy sense, what usually happens in plane crashes? Unfortunately, the answer is something called behavioral inaction or, in aviation parlance, negative panic. It occurs when people do nothing to save themselves, like the statues in the last chapter. In nature, when baby chicks freeze as the shadow of a hawk crosses the ground, this response is called tonic immobility. Humans don't have the same reflex, but the experience is similar. The current theory of behavioral inaction goes like this: As your frontal lobes process the sight of an airplane wing on fire, they seek to match the information with memories of similar situations in the past. If you have no stored experience of a plane crash, your brain can't find a match and gets stuck in a loop of trying and failing to come up with the right response. Hence: immobility. The military calls this process the dislocation of expectation. On airplanes, this freezing reaction is "less spectacular, but probably more common and dangerous than panic flight," according to Daniel Johnson, a well-respected aviation safety expert. When a passenger sits motionless on a burning airplane, Johnson writes, he isn't "unaware of the danger. On the contrary, he knows that if nothing is done, severe pain and even death will probably occur—but still he does nothing." Johnson believes this response has less to do with fear and confusion and more to do with "the novelty of the situation and the lack of leadership." People simply don't know how to respond when something unexpected happens.

It doesn't match their experience or expectations. So they do nothing. They wait for instructions. And they often die.

Is it possible to tell if you're prone to behavioral inaction? Is there anything you can do to reduce the likelihood? And perhaps most important, what additional steps can you take to improve your chances of survival in a crash? The answers to those questions make up the rest of this chapter.

3. What Your Flight Attendant Really Thinks of You

When you board a plane, you're probably accustomed to some friendly banter from the flight attendants. They welcome you, ask where you're from, and make idle talk about the weather. It may seem like empty chatter, and you may wonder how they can stand saying "Hi, how are you?" to three hundred travelers. But this is serious business. They're not just being friendly; they're profiling you. For starters, they're checking to see whether you're fit for flying or whether you're under the influence. They're also looking for suspicious behavior including clues of terrorist activity. One of their other main objectives is to identify ABPs. In the parlance of flight safety, that means "able-bodied passenger." In an emergency, ABPs are the ones they call upon for help. Flight attendants are trained to identify ABPs as they board and to keep track of where they are on the plane. ABPs are typically solo travelers. They're alert, healthy, and physically fit. They're often wearing clothing that suggests some kind of military, law enforcement, or firefighter training. They're likely to be in the top 10 percent on John Leach's 10–80–10 scale.

It may surprise you, but a flight attendant's real job isn't to serve you a Coke or fetch a pillow. They're on the plane to keep you safe. From the second you set foot on board, they're watching you, trying to figure out if you'll help in a crisis or if you'll pose a problem. They're checking to see if you're focused on your safety information card or if you listen during their briefing. That's another sign of an ABP—someone who shows an interest in surviving a plane crash. No surprise, most people don't.

Cynthia Corbett works as a principal investigator in cabin safety research at the Federal Aviation Administration. She's energetic and outgoing; you wouldn't immediately peg her as a statistician. She's got a stylish haircut, chic eyeglasses, and a purple manicure. Trained in experimental psychology, Corbett is a specialist in human factors in airplane crashes. She's also a safety zealot. She's crunched the numbers on who gets out alive and who doesn't. When I ask her the bottom line—who survives?—her answer is blunt: "Young, slender men." Agility and strength make the biggest difference when you're trying to wriggle through airplane wreckage or slip through a twenty-inch-wide emergency exit. In the most extensive study ever conducted on the subject, FAA investigators put 2,352 people through simulated evacuations. These were inexperienced folks—"naive" in the FAA's parlance. For extra motivation, they offered to pay some people twenty-two dollars—twice the hourly rate—if they were among the first 25 percent to get out. Turns out that when it comes to escaping from a plane, many of the things that you think might matter really don't. The width of the aisle doesn't have much effect. Nor does the motivation of the extra money. "It's the people who are the issue," says Corbett. Thirty-one percent of the differences among people's evacuation times depends on their personal characteristics. The biggest factors are age, gender, and what the FAA delicately calls "girth." In other words, if you're older, female, and large, you may have a harder time escaping a burning plane. That doesn't mean you're doomed—in fact, the FAA notes that one subject who weighed 416 pounds managed to get through a small over-wing exit with minimal difficulty and average speed. But the bottom line is that older, bigger women are usually less nimble and strong. In a crisis, they face the toughest challenge. Those three factors—old, female, and wide—the FAA concluded were "particularly harmful vis-à-vis evacuation performance."*

Once you allow for the personal differences among passengers, there's an even more important factor to getting out alive. Between 35 and 40 percent of the difference in evacuation times was explained by the effects of naïveté, in other words, inexperience and a lack of familiarity at fleeing planes. The FAA's conclusion was simple and straightforward. The biggest problem in cabin safety is "a failure of passengers to understand and properly execute emergency procedures." Again, the FAA has carefully analyzed the reasons. It turns out that 61 percent of fliers say they don't bother to listen to the safety briefings or read the cards. The title of the FAA's blistering report on passenger safety awareness says it all: *Still Ignorant After All These Years*. And who are the most inattentive and least informed fliers? Surprise. Frequent fliers are the worst. "They think they already know it all," Cynthia Corbett says. "They are a bit arrogant." When they get aboard, they believe it's more important to read *The Wall Street Journal,* make phone calls, or send e-mail. New fliers are the most attentive. They listen to the briefings and read the cards.

But here's the biggest and most vexing problem of all. Whether or not you pay attention doesn't seem to make a difference in your knowledge level. The FAA found that both attenders and nonattenders—people who reported that they paid attention or not—scored just as poorly in their awareness of what to do in an emergency. Whether they listen or not, they don't seem to internalize the relevant information. For instance, most passen-

---

* In FAA exercises, short-legged women and men over six feet tall also struggled with the over-wing exits that are narrow and require a step up to escape.

gers believe you can survive an hour without an oxygen mask after a plane decompresses at high altitude. In fact, you've got only a few seconds. They also believe you've got thirty minutes to flee a burning plane. In reality, as you're about to see, you've got only ninety seconds.

The first-class cabin was quiet and almost empty as David Koch stretched out in seat 2A. The journey from Columbus, Ohio, had been relaxing, and he was looking forward to a quiet evening in Santa Monica. From his window, he could see the orange sun setting over the Pacific Ocean. His shoes were off, his coat was folded on the seat next to him, and he noticed a pleasant older couple across the aisle. Just after 6 PM, the jet with eighty-three passengers touched down on Runway 24 Left. Koch remembers the familiar screech of the wheels and, a few seconds later, "a sudden, sickening crunch." A spray of sparks and a ball of fire lit up the window. He was sure they had struck another airplane. The cabin echoed with screams as a flight attendant shouted: "Stay down! Stay down! Stay down!" Koch quickly unbuckled his seat belt and got ready to run for the exit, but the plane kept skidding. Around twenty seconds later—what seemed like forever—it finally slammed to a stop, and another explosion shook the plane. Without a seat belt to hold him down, Koch was thrown forward into the first-row seat and then against the bulkhead.

Fear swept the plane as the cabin lights went out. Koch watched passengers run down the aisles toward the rear of the plane. Thick, choking smoke filled the cabin. The intercom shut off. The flight attendants offered no more evacuation instructions. The passengers were on their own. "I immediately got on my hands and knees and attempted to find my shoes," Koch remembers. "I believed it would be difficult to escape a burning plane in my stocking feet." But he had no luck with his shoes or coat, which he had hoped to use as a mask.

Crawling toward the back of the plane, he searched for an exit. Koch is six foot five and an imposing figure, but his size didn't matter on the floor as other passengers stampeded over him. "After moving a few rows," he remembers, "I encountered a fighting, frenzied mob jamming the aisle." The congestion, he says, "suddenly made me realize that escape was probably impossible because I was last in line to get out the rear exit. I concluded that I was probably going to die. At that point I stood up and, choking heavily on the smoke, walked back toward the first-class section."

In the midst of chaos, Koch's thoughts were clear and calm. "I was not panicked nor was I terrified." He simply couldn't believe that he was fighting for his life when just a few moments earlier he had been reading happily and looking forward to an evening in Southern California. "For a few long moments I stood there," he says, "immobilized, not knowing what else to do, and I knew with absolute certainty that I was going to die." Then Koch describes an amazing sensation. He felt his mind separate from his body and rise up toward a white light. He remembers looking back at his body dying in the airplane while his mind successfully escaped. Suddenly, he continues, his brain snapped back into his body and he realized there had to be a way out. If smoke was pouring into the front of the plane, there had to be an opening in the fuselage that might also offer a way to get out. He stumbled forward toward the cockpit. When he reached the passenger door, he could see a roaring fire through the circular window. Opening the exit would have meant instant incineration. "I was totally alone but not panicked or terrified," he remembers. He turned around and stepped to the opposite wall. He was feeling faint—he guessed later that he had only ten to fifteen seconds of consciousness left, because every breath of black smoke was tremendously painful. Then he saw a crack in the fuselage. He pried his fingers into it and pulled hard, realizing it was the galley door. He later learned that the steward had opened it right after the crash and then tried to shut it—as he was trained to do—because of fire and smoke on the other side. Koch shoved his head outside and gulped down a few breaths of air. "A tremendous sense of strength came over me," he says, "and a wave of adrenaline shot through my body." For a moment, he felt as powerful as Superman.

Below him, he could see fire burning under the plane. *Oh what the hell,* he thought, jumping past the flames onto the asphalt ten feet below. He landed hard, and crawled away from the plane. When he looked back, "the sight was a nightmare." He watched passengers struggling and squeezing to get through the exits. On the ground around him, other survivors seemed dazed. Some were confused and silent, staring blankly. Some shook and sobbed. Some were covered head-to-toe in white foam injected into the cabin by the fire crews. "I consider it a miracle that I escaped," he says, "and that I came through the ordeal as well as I did."*

The date was February 1, 1991. As USAir 1493 touched down, it collided with SkyWest 5569, a twin-engine commuter plane waiting for takeoff with ten passengers and two crew. A confused and distracted air traffic controller had mistakenly cleared the 737 to land on the same runway where one minute earlier she had sent the smaller plane. The USAir jet hit the SkyWest plane at 150 miles per hour and the two aircraft careened across the airfield, smashed into an abandoned fire station, and burst into flames. Everyone on the SkyWest turboprop died.

* Today Koch and his brother Charles preside over Koch Industries, America's largest privately owned company, with interests in manufacturing, energy, and commodities. He is reportedly worth seventeen billion dollars, and *Forbes* magazine ranked him as the thirty-seventh richest man in the world in 2008.

Twenty-two people on the USAir jet were killed, including eighteen who couldn't get off the plane. When the fires were put out, the charred bodies of the first-class flight attendant and ten passengers were discovered lined up in the aisle less than eight feet from the over-wing exit. The official investigation concluded: "They most likely collapsed while waiting to climb out." This was the pileup that David Koch encountered. If he had not turned around, he would surely have been the eleventh passenger to perish in that aisle.

Safety board investigators interviewed survivors and found a number of disturbing facts about the evacuation. A woman seated in 10F, the emergency row, admitted that she froze and was unable to leave her seat or open the window exit next to her. A male passenger in 11D climbed over a seat, opened the hatch, and pushed the woman out onto the wing. A few moments later, two male passengers "had an altercation" about who would evacuate first. The fight lasted several seconds. These two events "significantly hampered the evacuation," the safety board concluded. In short, passengers died on USAir because one woman experienced behavioral inaction, blocking the exit row, and two men fought to get out the door. By some estimates, up to 30 percent of fatalities in airplane crashes fall into this preventable category. But why is it so difficult to escape? What really happens when smoke fills the cabin? To understand, I traveled to Oklahoma City, where the Federal Aviation Administration teaches its master class in how to evacuate from a burning plane.

4. Ninety Seconds to Save Your Life

The first thing you notice is how fast you get lost. I'm supposed to be following a line of fellow passengers through the cabin toward an exit and yet I can't see a thing, not even the Canadian woman who I know is a couple of steps ahead of me. Even trying to follow her voice is not easy. The smoke engulfs and overwhelms you. This may be just a drill, but it messes with your mind. Even though I'm surrounded by twenty-eight other people, I feel alone stumbling through the murk.

Four times a year near Will Rogers World Airport, at the red-brick headquarters of the Civil Aerospace Medical Institute, the FAA holds a three-day workshop on surviving plane crashes. Renowned around the globe, the classes are taught by some of the leading experts in aviation safety. In my class, the other participants are safety managers and in-flight instructors with airlines like American, Comair, Horizon, Northwest, SkyWest, and USAir. Three men are here from Korean Air and two women from Air Canada. I'm the only outsider. We've all come to learn the latest on how to herd passengers off smoky planes and how to survive crash landing in the ocean. Above all, we're here to confront the universal challenge of aviation safety: What can be done about the vast majority of fliers who aren't ready to survive a crash?

To get us in the mood, the trainers play the stunning amateur video of the crash of United 232, the one that Jerry Schemmel survived. The horrifying scene is one of the most widely watched images in the history of television. Shot through a chain-link fence, the tape captures Flight 232 landing hard, exploding, and careening down the runway. Even in this classroom of safety professionals, there's a hush as the fiery images play on the screen. It's hard to believe anyone made it out alive. Then they take us to the cabin evacuation simulator inside the chopped-up fuselage of a C-124 Globemaster cargo plane. The nose and tail have been lopped off, leaving a hulking cylinder on huge hydraulic lifts that jut from a green lawn next to the FAA building. It's a hot, sunny day but inside the simulator, it looks and feels like an air-conditioned jetliner. We each take a seat and buckle up. We're given our instructions—we'll evacuate through a floor exit at the front of the plane. Then two machines start pumping nontoxic smoke—the kind used in theaters and discos. Poof. Within thirty seconds, we can't see a thing. A buzzer sounds—someone shouts "Evacuate"—and it's time to move.

Despite the seriousness of our business, the mood is light. "I have a baby, let me through!" one classmate jokes. She's carrying a ten-pound infant dummy that's used in crash tests. As another woman jumps down the evacuation slide, she turns to her friend and asks: "Any last words?" But it's not all fun and games. One student, a sixty-two-year-old woman in charge of safety instruction for a small airline, wipes out at the bottom of the slide and breaks her ankle. She's rolled away on a wheelchair. Later, a trainer whispers that this injury is proof that older women are especially vulnerable in evacuations.

I stumble through the smoke, trying to stay low where the air is supposedly clearer. I'm so focused on following the track lights on the floor that I bang my head into an armrest and carom off a bulkhead. Later, I poke an eye on something that's sticking out in the haze. I see the glowing light of the doorway and press toward it. Suddenly the air clears and I'm standing at the exit, squinting into bright sunlight, staring down at the nine-and-a-half-foot drop. The gray emergency slide beckons—its slope is thirty-six degrees and it looks surprisingly steep.* I "jump and sit," whoosh, and it's a very quick ride to the bottom. I'm next to last off the plane. On the ground, I check my watch. It took two minutes and twenty-five seconds to get out. We congratulate ourselves. "That was fun! Let's do it again," a woman says. But our instructor isn't impressed. Two minutes and twenty-

* When the FAA tested slides with forty-five-degree angles, passengers came down too fast and hurt themselves. When they tried forty-eight degrees, no one would jump because the angle looked too dangerous.

five seconds is far too slow. In a real emergency, we would be dead.

The rules of survival in a plane crash start with the fact mentioned above: You have only ninety seconds to get out. That's all. Ninety. Any longer and a fire could burn through the aluminum skin of the plane and the cabin temperature will soar to more than two thousand degrees. Soon after, a flashover fire will consume everything. In just ninety seconds, the cabin turns into an inferno. That's less than the time it probably took to read this page. The next thing to know is a concept called Plus Three/Minus Eight. In aviation parlance, *Plus Three* refers to the first three minutes of flight, and *Minus Eight* to the last eight. Plus Three/Minus Eight is prime time for an emergency. Flight attendants at one major airline are taught that 80 percent of all plane crashes take place during those eleven minutes. Wet and icy runways are the leading cause. Now that you know the principle, what can you do with this knowledge? David Palmerton, the FAA's resident expert on crashes and protection, says the key is vigilance during Plus Three/Minus Eight. At the start and end of a flight, the impact forces are typically less severe because a plane's speed is slower and its altitude is lower. That means if you're paying attention, there's more opportunity to survive. You should never drink a beer or a martini before getting on a plane. You should never pop a sleeping pill before flying. You shouldn't nap or listen to your iPod when you're rolling down the runway. And you definitely shouldn't wait for the thud of landing to wake up. In the first three minutes and the last eight, you should be ready to run for your life without waiting for orders or instructions. According to one study, 45 percent of the flight attendants in survivable crashes are incapacitated. That means almost half of the time, you're on your own.

From the moment you step onto the plane, you should begin

formulating your action plan. The key word is *action,* Cynthia Corbett says. Figure out what you're really going to do if you need to escape. Of course, this idea runs contrary to the entire experience of going to the airport and flying these days. Every step of the way, from ticketing to security to boarding, you're told exactly what to do and when to do it. Too many passengers carry that passive mentality into a crash.

If survival is entirely up to you, what are the most important things to know? Above all, memorize where the emergency exits are located and count how many rows away they are. In thick smoke, you may not be able to see where you're going, but you can feel your way by counting the rows with your hands. Knowing the location of one exit isn't enough. You also need a backup. Plan A—your first choice—may be broken, blocked, or on fire. Again, count the rows to Plan B—your alternative exit. During Plus Three/Minus Eight, focus on how you would get to your exits if the passageways were blocked.

Nora Marshall traveled for sixteen years as a flight attendant with World Airways before joining the National Transportation Safety Board. Today she's chief of the NTSB's Survival Factors Division and has spent the last twenty-five years examining who lives and dies in plane crashes. Some factors are beyond your control, she says. For instance, are the crash forces within human tolerance levels; does the structure of the cabin remain "substantially intact"; and is there enough "livable volume" for the passengers? Other factors, however, are up to you. Is your seat belt properly fastened; does the restraint system hold you in place; and do you evacuate fast enough? Before every takeoff and landing, Marshall closes her book or magazine and reviews her action plan. After so many years as a flight attendant, the routine is an "an unconscious habit." She's aware of all the exits. She buckles her belt low and tight across her hips. She's pays attention to what's happening. She's ready to go.

Anticipating what you'll do in an emergency is absolutely critical. "Even rats will perform better if they know what's expected of them," says the FAA's Cynthia Corbett. "It's true." In a crash, a few seconds of hesitation can make the difference between life and death. What's more, without an action plan, the most unfortunate things happen. Husbands abandon their wives. Parents flee without their children. At the very least, you should remind yourself on takeoff and landing who matters most to you. You might even repeat this mantra: *I've got kids! I've got kids!* And if you're flying with children, Corbett says, you should discuss an escape plan even before you get on the plane. In a crisis, you might not have time to search for them, and just like at home, they should be "trained" for emergencies.

The fundamental question to ask, Corbett says, is: How much is your life really worth to you? Some evacuations are civilized and orderly, like the miracle of Air France Flight 358. On August 2, 2005, the Airbus A340 landed in Toronto in gusting winds and heavy rain. The plane overshot Runway 24 Left and plunged into a ravine before bursting into flames. Fire swept through the middle of the plane, blocking some exits, but all 309 people on board escaped in ninety seconds. Emergency crews arrived at the scene within fifty-two seconds. It was a model evacuation. Unfortunately, that's not always how it goes. This raises the question: Are you prepared to climb over seats and force your way past people frozen in fear or blocking your way? You'd better believe others will claw for the exits. David Koch was trampled while crawling to escape the USAir crash. The FAA has a gentle word for this: They call it "competitive" behavior. Are you ready to go for it?

The final step in your action plan involves understanding what actually happens to your body in a crash. Rick DeWeese is an engineer who runs the FAA's Biodynamics Lab in Oklahoma City. On a tour of his workshop, he introduces me to various crash dummies that he uses on a test track to simulate what happens to our bodies when airliner seats abruptly come to a stop. Then he shows me videos of mock crashes with his dummies violently whipsawing at the waist. Even with your seat belt properly fastened, he explains, you're still going to smash into something. That's intentional. Indeed, the seat in front of you is actually part of your safety system. In a crash, it's designed to catch you and slow you to an acceptable rate of speed. It's even got a special hinge that controls the deceleration. So face it, in an accident, you're going to hit something. DeWeese says it's a good idea to avoid bulkhead seats because those walls don't give way when you strike them. Also, it's important to understand that the brace position can really save your life. As safety information cards illustrate, this crash position involves leaning forward, placing your head on or near the surface it's most likely to hit, and putting your feet flat on the floor. This technique is designed to minimize the force of impact and so-called flail of your limbs when you're whipped or jackknifed forward. In particular, the brace position reduces the velocity of your head when it slams into the seat or bulkhead in front of you. And don't forget to remove all slack from your seat belt. It should be fastened firmly across your hips, comfortable but still tight. It's designed to withstand 3,000 pounds of force—that's seventeen g's for a typical 170-pound man. "People can survive these kinds of forces if they're restrained well," DeWeese says.

The experts rattle off a few other safety tips. If you want to increase your chances, you might consider bringing along something called a smoke hood, a device with a breathing filter to protect against toxic gases, irritants, and heat. Some hoods come folded up in plastic containers not much bigger than a cigarette pack, but there's real controversy about their effectiveness and desirability, especially because very few protect against carbon monoxide, a major chemical killer in airplane crashes and home fires. The best hoods are often bulky and expensive, but if sealed correctly around your neck, they may give you more time to escape a fire. On the flip side, they may also endanger you if they aren't easy to don and therefore slow you down, if they make you complacent, or if they impede your ability to hear safety instructions from the crew. Still, David Koch, who survived the fiery crash of USAir 1493, always travels with a smoke hood in his briefcase, and one of the world's top fire safety experts recommends that you carry a quality hood whenever you fly or stay in hotels, and that you also keep one by your bedside at home.

Another lifesaving tip: In a crash, forget about your carry-on luggage. Don't even think about trying to rescue your laptop, your iPod, or your Dan Brown novel. Too many people perish because they try to save their toilet kits. Lugging your bags will slow down your escape and block others, too. In terms of clothing, the new security regulations make it tempting to wear sandals or slip-ons. They're more convenient for the TSA screening line, but in an emergency you're going to want solid shoes—preferably lace-ups—that won't fall off. Running barefoot through broken glass and burning metal isn't fun. Women should travel in flats—no high heels—and they shouldn't wear stockings or synthetic fabrics that can melt on the skin. Short pants or skirts are another no-no. In a fire, you'll want your body covered.

If you take all this seriously, there's one last thing you can do. When she gets on a plane, Cynthia Corbett makes a big fuss about reviewing the safety card. Her research shows that people are more apt to pay attention if others are doing so. She talks loudly about the nearest exits and asks her neighbors to explain the safety instructions. Her goal is to influence the people around her to focus. It may save all of their lives. She doesn't want anyone nearby to freeze and perhaps block her way.

In the end, crash survival comes down to a simple question, Corbett says: "How committed are you?" Your mind-set is the key. It would be wonderful if every time you got on a plane, you found a cabin filled with young, slender, healthy men or ABPs. Alas, that's never going to happen. In the unlikely event of a crash, it's going to be up to you to get out alive. Survival of the fittest doesn't just mean physical fitness, Corbett says. It means mental fitness. Memorizing the safety card and your evacuation plan can significantly improve your chances, even if you're large or less agile. When Corbett travels, she always notices passengers who ignore the safety briefings, wear the wrong clothes, and treat the flight like some kind of party in the air. "What is your life worth?" she wonders. Then she thinks: "I'm going to survive and you're not."

5. The Safest Seat on the Plane

I've always believed the safest seat on a plane is in the very back. It seems logical. If a plane crashes nose first, you're last to arrive at the accident. In July 2007, *Popular Mechanics* bolstered this view when it published an analysis of every commercial jet crash in the United States going back to 1971. Comparing seating charts and survival rates, the magazine concluded: "It's safer in the back." Examining thirty-six years of real-world accidents, the author claimed that "the farther back you sit, the better your odds of survival." If you're flying near the tail of a plane, he explained, you're 40 percent more likely to survive than passengers in front rows. Specifically, the rear cabin had the best survival rate—69 percent—compared with 56 percent for the over-wing and coach sections. First and business class had the worst rate, 49 percent.

When I ask government experts about these conclusions, they politely but emphatically disagree. All aviation accidents are different, they say, and there's simply *no* proof that one particular section is safer than another. Rick DeWeese of the FAA explains that it's impossible to pick the safest seat because you never know what kind of accident you'll encounter. Some planes crash tail-first. Others hit wing-first. Regardless, survivors *and* victims are often found right next to each other in the same sections of the plane. That said, DeWeese always tries to fly on bigger planes because they have more energy absorption or "crushability" in a crash. That means your body is subjected to fewer deadly forces.

So where should you sit next time you fly? Forget about the front or the back—the statistics are inconclusive, if not maddeningly contradictory—and just remember the Five Row Rule.* Professor Ed Galea of the University of Greenwich in London is a leading authority on fire safety who has carefully studied 105 plane crashes, seating charts, and the accounts of 1,917 survivors and 155 cabin crew members. He discovered that survivors usually move an average of *five* rows before they escape a burning aircraft. That's the typical "cutoff," he tells me, even though some survivors have managed to scamper as many as nineteen rows. Perhaps it's predictable, but Galea argues that passengers are "most likely to survive" if they're sitting right next to the exit or one row away. If you're between two and five rows from an exit, while you're more likely to live than die, "the difference between surviving and perishing is greatly reduced." Beyond five rows, Galea warns, "the chances of perishing far outweigh those of surviving." Hence his recommendation: Always sit within five rows of an airplane door. Of course, there's

---

* The latest research from the University of Greenwich directly contradicts the *Popular Mechanics* findings: Passengers at the front have a 65 percent chance of survival compared to 53 percent in the rear.

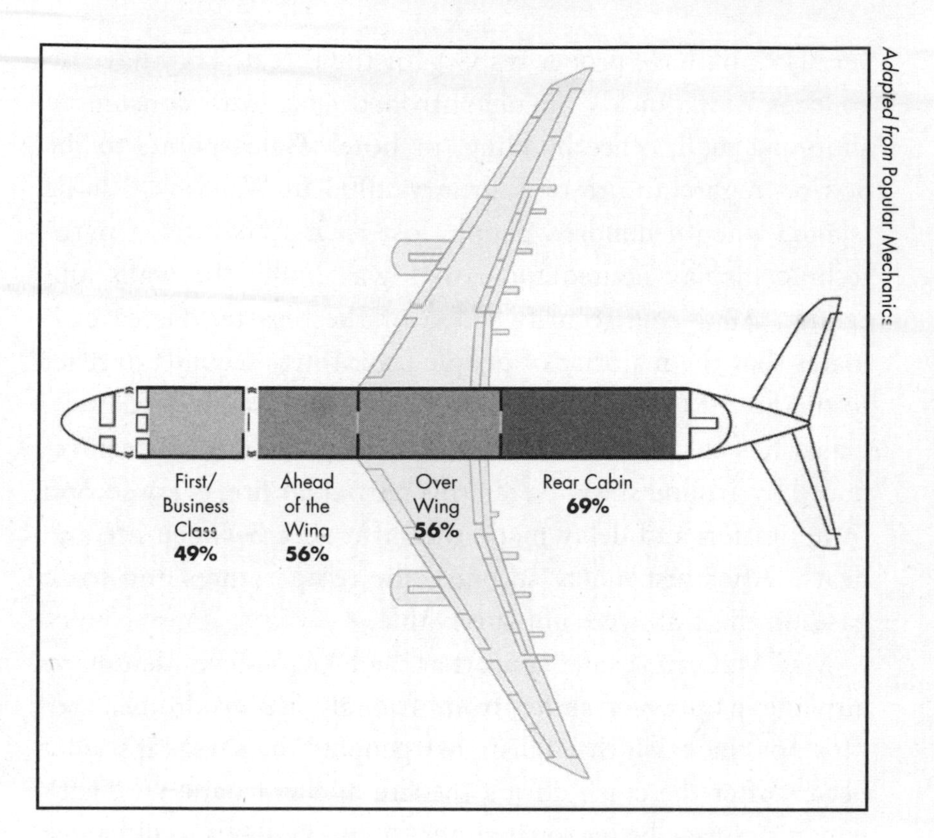

The *Popular Mechanics* analysis of survival rates on passenger planes.

no guarantee the exits will be "viable" in an accident. A door could be jammed, destroyed, or aflame. That's why he also suggests aisle seats for "marginally" more mobility and options. Passengers in the aisles have a 64 percent chance of getting out alive compared to 58 percent by the windows.

After studying all kinds of infernos—including blazes in department stores and nightclubs—Galea worries that most people don't recognize the danger of fire. Indeed, many suffer from what he calls friendly-fire syndrome. Their only direct experience involves roasting hot dogs or lighting logs and they simply don't have proper respect for the threat. They have no concept how quickly an uncontrolled spark can consume a shopping mall, office building, or hotel. Galea points to the Station nightclub fire of February 2003 in Warwick, Rhode Island, when a hundred people lost their lives after a pyrotechnic display ignited foam that was lining the walls and ceiling. Analyzing security video of the disaster, Galea estimates that the majority of people took thirty seconds to react to the fire. If they had responded and evacuated immediately, Galea has calculated that the death toll would have been reduced by around sixty-five. In this particular fire, every second of hesitation and delay made the difference between life and death. After just ninety seconds, the temperatures and toxic gases in the club were not survivable.

Mac McLean, a safety expert at the FAA, believes that many airplane passengers suffer from friendly-fire syndrome, too. "It's not the crash that kills most people," he says. "It's what occurs after the crash during the fire and evacuation." That's why it's always better to sit closer to an exit—but if, like most budget travelers, you don't have a choice, you shouldn't worry. No matter where you're located, it's most important to prepare yourself with an escape plan and a backup. As for the special challenges of older, wider women, the answer is simple. Avoid the middle seats of the plane and the over-wing exits where you would have to squeeze through a narrow portal. Try to sit near the so-called floor exits—the bigger and wider main doors, usually at the front and back of the planes.

Arnold Barnett, the MIT statistician, offers one final way of thinking about the best seat on the plane. Despite the incredible safety of air travel, if you're still worried about flying, try upgrading your ticket to get a more comfortable seat, with more pampering from the flight attendants. If you can find a way to reduce your anxiety level, you'll actually be safer. After all, your risk of a stress-induced heart attack in the air is far greater than your chance of dying in a crash.* So when you make your next trip, try to sit within five rows of an exit. Memorize an escape plan. And perhaps most important of all: try to relax.

* In a similar vein, extreme fear of terrorism may be more hazardous to your health than al-Qaeda itself. People who were acutely stressed after September 11 had 53 percent more cardiovascular ailments over the next three years, according to a recent study from researchers at the University of California–Irvine. If only 0.0003 percent consequently died from such stress-related heart problems, "that would be higher than the 9/11 death toll," according to an estimate in *The New York Times*.

SURVIVAL SECRETS

The Woman Who Fell from the Sky

Vesna Vulović's first trip to Czechoslovakia broke all the rules. There was no passport inspection at the border; no first step on Czech soil; no welcome sign to greet her. Vesna's arrival was like no other in history. "I fell from the sky," she told the *Prague Post*. "I came without a visa." At 5:05 PM on January 26, 1972, Vesna was standing in the aisle of JAT Flight 364 heading from Copenhagen to Belgrade. A brand-new flight attendant, the blond twenty-two-year-old was handing out meals when a bomb hidden in the luggage exploded, ripping the DC-9 apart. Everyone died—twenty-seven victims of fire, rapid decompression, freezing temperatures, and high-speed impact with the ground.

Impossibly, Vesna plummeted six miles—thirty-three thousand feet—and lived. To this day, her name appears in *Guinness World Records* for surviving the longest fall without a parachute. After crashing to earth in northern Bohemia, near the German border, she was unconscious with severe head injuries. Her spine, legs, pelvis, and ribs were broken. The midair explosion shook the ground in Srbská Kamenice, a small town where villagers saw the falling debris. A local man heard cries in the wreckage. "He saw two legs sticking out of the fuselage," Vesna told *Life* magazine. "It was me." The man was trained as a nurse and was careful not to move her. She was wearing her turquoise airline uniform, and he covered her with a coat and cleared her airway until medics arrived. In a coma for three days, unable to move from the waist down, Vesna awoke in a small hospital and said in perfect English: "Can I have a cigarette?" Then she asked her mother: "Where are my cats and dogs?" Within a year, she was walking and had returned to a desk job. The secrets of her amazing recovery, she told *The New York Times*, included eating chocolate, spinach, and fish oil as a child and her innate Serbian stubbornness. After nearly two more decades on the job, she retired to a Belgrade apartment crammed with religious icons. Her spine is still crooked and she limps when she walks, but she is proud of her Guiness record. "If you can survive what I survived," she says, "you can survive anything."

Dr. Richard Snyder knows more than anyone in the world about free falling and "human impact tolerances," the technical phrase for how much the body can withstand. Over a fifty-year career, he has researched more than thirty-three thousand falls of every height and variety. As a crash injury expert at the FAA in 1963, he published a classic study of 137 falls, including a sixty-nine-year-old woman who toppled from a tree while chasing her pet parakeet and an eloper who tumbled from a tall ladder. Snyder's subjects ranged in age from eighteen months to ninety-one years old. Humans, he concluded, are able to survive impact forces "considerably greater than those previously believed tolerable."

Snyder is eighty now and retired in Tucson, Arizona, after careers in government, academia, and the private sector. He tells me he has always been the most fascinated by people who fall from planes. He collected a thick file on Vesna Vulović. He also gathered information on more than a thousand people who survived falls *greater* than a thousand feet. It may surprise you, he says, but "it isn't that unusual." It all depends on how you fall, where you land, and your own physical condition. Snyder's early research found that "psychotic patients," like suicidal people, may be able to withstand impact forces better than "normal" individuals. A crazy person who wants to leap from a building, he theorized, may be looser on the way down. "The act of jumping may thus be a release for him," he writes, "and unlike most of us, this individual may enjoy the jump. As a result he may be physically relaxed at the time of impact, which appears to be, in itself, an important criterion for survival of free-fall." Snyder discovered that people who are drunk also "appear to have a disproportionate survival rate among free-falls of extreme distances" because they were "abnormally relaxed."*

Snyder rattles off stories of free fallers the way some tell tales of great sports legends. He recalls that one man lived for ten hours after falling thirty-nine thousand feet when a Boeing 707 exploded. Had the victim received proper medical attention, he would have survived. A Russian pilot fell twenty-eight thousand feet and landed in deep snow on the side of a mountain, where he was rescued by Cossack horsemen. Snyder wrote a whole paper on snow as an "impact attenuator." He says if you happen to fall a great distance, pray that you land in what's known as corn snow, which has large, round crystals from repeated thawing and freezing.

Snyder is proud of his fifty years exploring impact forces. His work has helped designers make safety improvements in planes and cars that protect you every day. But he recognizes there's much more work to do. On the one hand, people like Vesna survive thirty-three-thousand-foot falls. "On the other hand," he says, "you can trip and fall going out the door and die." He chuckles. "It's kind of an ironical business."

---

*For more on who lives and who dies in free-falls, please see Appendix A, "The Science of Falling Cats (and Babies)," on page 341.

4

The Organ Recital

Who Lives and Dies in the ER

When the unknown woman arrived on the roof at Stanford Medical Center, rushed there by helicopter ambulance, she was identified only as Twenty-two India #99330292. A flight nurse gave her this patient ID, the kind you get when you're scooped up from the streets after a terrible accident, you're unable to offer your name, and you don't have any identification. Twenty-two India had already flatlined once at a smaller hospital where the first chopper had flown her. When her heart stopped beating, doctors shocked it back into rhythm, but they were not equipped to handle all of her injuries. So another copter carried her to the bigger trauma center at Stanford, where she went into cardiac arrest two more times in the emergency room.

The facts of her case were painfully straightforward. Riding her bike on a two-lane road, Twenty-two India had been blindsided by a forty-three-thousand-pound Volvo truck. The scans showed a woman broken in almost every possible place: neck, clavicle, scapula, sacrum, all right-side ribs, sternum, patellas, right tibia, right fibula, and pelvis. All told, more than twenty bones were damaged—but they would heal, so ER doctors focused on her real problems. The collision with the twenty-one-ton truck had transected or severed her aorta, the same kind of injury that killed Princess Diana. In addition, Twenty-two India's lungs collapsed, her liver was shoved into her chest cavity, her diaphragm was ruptured, and her kidneys were bruised. Later, when the trauma doctors calculated her probability of survival, a statistic used by surgeons to review their cases, the result was sobering. Twenty-two India's chance of living was zero.

We drive along a country road in the Livermore Valley, just over the hills from San Francisco. Vineyards stretch in every direction under the caramel light of morning. This is the hour of the accident. The woman once known as Twenty-two India sits beside me. Her name is Katharine Decker Johnson. She's fifty-seven years old, and everything about her seems elongated and vertical, like some kind of wading bird. Her blond hair plunges straight to the middle of her back; and her bangs push down in front, shielding blue-green eyes. She's five foot eleven with slender legs that barely fill black stretch pants. She weighs 118 pounds now, down from 138 before the crash. Growing up on a five-hundred-acre farm in Ohio, she preferred the company of horses to humans. "I was afraid of everybody," she explains. Always happiest when she was alone and outdoors, her encounter with the twenty-one-ton Volvo truck has changed her in many ways. Since the accident, she feels more connected to people.

On the outskirts of Livermore where the housing developments stop and the grapevines start, we visit the spot where she was hit. It's an ordinary three-way intersection that funnels commuters toward the entrance of the national laboratory a few miles away. At 8 AM on October 27, 2003, Katharine was training on a new bike for a hundred-kilometer race. "Kapow!" she says. The truck smashed into her, and she bounced right off, leaving a body-size dent on the driver's side of the slanted hood. Fortunately, paramedics gave her some "milk of amnesia," she says, so like many trauma victims she doesn't remember anything about the impact. She points to the valley oak tree that stands in an empty field beside South Vasco and Tesla roads. "My tree," she calls it. That's where the first helicopter landed to ferry her to the hospital. She comes here twice a week to walk and meditate. "I always pray when I'm here," she explains. "I always say the 'Glory Be.'" She smiles for a moment. "Glory Be that I'm alive. There's more work for me to do."

© Gary W. Johnson

Katharine Decker Johnson's bicycle.

Over breakfast at the Railroad Café, a local joint where ceiling fans whisk the air, she shows me a scrapbook of her ordeal. Katharine is a scientific illustrator, and she has drawn detailed pictures of all her broken bones. In one image, she's on horseback, and you can see the five places where her pelvis shattered. Snapshots in the album capture the ordeal from moment of impact to the present. Her mangled bicycle. So many bandages in the hospital. A faint smile as she sits up in bed. Doctors. Nurses. A guitar-playing therapist. Then she tells me about each of her injuries, a litany that she calls her "organ recital." So much damage in so many places. After eight months, when she finally stood up for the first time, she needed an external fixator, a metal device like an Erector set to stabilize her healing bones. Then she learned to walk all over again. "I'm in pain all the time," she says. "There's something always out of whack." And yet as we part ways, she tells me she's going for a morning ride—and some jumping—on a horse named Carlos. It's perfect therapy, she says. When I ask incredulously about the risk to her mending bones, a big smile says don't worry. She's got a custom-made saddle especially cushioned for her "creaky body."*

1. The Most Unusual Reunion of the Year

Every year across America, millions of people gather for reunions. Families assemble on back porches for barbecues and tales about Grandma and Grandpa. High school classes get together to mark the passage of time and reminisce about senior prom. At the Arrillaga Alumni Center at Stanford University, they come together for what might be the most unusual reunion of all. It's a gathering of men and women like Katharine Decker Johnson who shouldn't really be alive, survivors of terrible traumas like head-on car collisions, motorcycle wipeouts, and even worse. Some travel hundreds of miles for the event. They roll in on wheelchairs and shuffle with canes. They wear protective helmets and braces. Some show no signs of their injuries, hiding their scars beneath their clothes. Around two hundred

* In January 2007, Katharine was bucked from her horse and fractured her right hip. After successful surgery, she's already back in the saddle. With admiration, her husband, Gary, says she's got "good bone."

people who gather to eat hors d'oeuvres, sip cold drinks, and swap stories. The able-bodied even dance to live music from a three-piece band. They want to rally around one another, embrace new arrivals who have recently been slammed by life, and show gratitude to the doctors, nurses, and social workers who put them back together. Above all, they're here to remind themselves they're still alive. It's called the Trauma Survivor Reunion, organized by the Stanford Medical Center. On a splendid day in May, I make my way through the crowd, catching fragments of conversation about painkillers and MRI's. I've come to meet the members of this remarkable survivors club and to learn who lives and dies in car wrecks, electrocutions, explosions, shootings, and stabbings.

It took rescuers more than forty minutes to pry Ricky Bunch from the cab of his Ford F150 pickup. Pinned upside down, he was in terrible shape when they found him at the bottom of the creek. Suspended for sixteen hours, he was unconscious and barely alive. The high school junior had just gotten his driver's license and a new truck, an early holiday present from his parents. He was a tough kid and competitive athlete who played linebacker—number 28—for his high school football team. He was also a blues guitarist, martial artist, and motorcycle buff. On Christmas Eve 2002, he was under the weather with the flu and didn't feel well enough to go to his grandmother's house for dinner. So he stayed in bed to rest. Sometime that night, he took a short drive to drop off a gift for his new girlfriend. Mount Eden Road winds through the hills, and on his way back he somehow missed a curve, drove across the oncoming lane, and flew off a cliff. His truck swiped two trees and crashed thirty feet below into Calabazas Creek. The cab landed upside down in the riverbed, and the driver's side was almost flattened. Weather reports had called for rain, and if the water had risen just an inch or two, Ricky surely would have drowned. But the skies never opened up. Sixteen desperate hours later, a friend spotted the truck. Investigators found no signs of alcohol or drugs and blamed the accident on Ricky's inexperience behind the wheel.

Courtesy of Dona Bunch

The driver's side of Ricky Bunch's Ford F150.

When the rescue helicopter brought Ricky to Stanford Medical Center, he was suffering from a crushed leg, collapsed lung, and kidney failure. Doctors believe his hypothermia slowed his heartbeat and helped save his life. But his brain was badly rattled in the crash. The technical term is *diffuse axonal injury* or DAI, a trauma that occurs when the head is shaken by acceleration or deceleration. The shearing forces tear the microscopic axons that allow the brain's neurons to communicate. The result is often coma and death. From the very beginning, Ricky's parents say they were told he would never be able to walk or talk again. If he survived, he would probably end up in a vegetative state. After two months in a coma, he was transferred to another hospital. His weight had dropped to eighty pounds.

* * *

Ricky weighs more than 160 pounds today and looks like a typical teenage rocker. He's five foot six with stringy black hair to his shoulders. He's got a scraggly soul patch, a goatee, a Fender T-shirt, and AC/DC sweatpants. He's a different person now, his mother, Dona, says, sitting at their kitchen table. He used to be a tough guy but now he's softer. He speaks gently, sometimes searching for words as he shows me pictures of the pulverized front seat of his truck. He pulls his shirt down to reveal the tracheotomy scar on his neck. He lifts his pant leg to expose the huge gashes where he's had fourteen operations. How did he overcome so many challenges? Ricky introduces me to two more theories of who lives and who dies. "The survival secret," he says, "was that I was sixteen years old. That's the magic age." When he awoke from his coma after two months, doctors told him that he was the perfect match for his injuries. Any younger, and he wouldn't have had the physical reserves to battle back. Any older, they said, and his brain wouldn't have been able to recover. Ricky says he also survived because of all the love that surrounded him. His family and school friends came to his bedside every day. They covered his room with posters that said YOU'RE A FIGHTER! and for six months, the big sign in front of Saratoga High School encouraged: GET WELL SOON RICKY.

Today Ricky runs, rides a bike, plays guitar, and teaches and judges jujitsu. He graduated with his high school class, and he's enrolled in math and writing courses at De Anza Junior College in Cupertino. Someday, he'd like to be a paramedic or firefighter and climb down into ravines and cut people out of car wrecks. His mother wonders if he'll ever recover enough to achieve that dream. But then she looks at her son—proof that anything is possible—and says, "He's our Christmas Miracle."

2. Knives, Bullets, and Brick Walls

Life is traumatic. There's no avoiding it. More than 115 million people visit an emergency room every year in the United States. That's 315,000 times per day or 13,125 per hour. Snap your fingers and three people have been rushed through the ER doors somewhere in America. The kinds and quantities of injuries barely change every year. Trauma is painfully predictable. Cars and guns are always at the top of the list. Every fourteen seconds, someone gets hurt in a traffic accident, and every twelve minutes someone dies. Every hour, eight people suffer a firearm injury and three people perish. Two children are treated for choking injuries every hour, and three die every week. Dog bites send forty-four people to the ER every hour. Nine people die every day from accidental drowning and almost three from electric shock.*

When it's your turn in the ER, how can you calculate your chances? It turns out there's a formula for predicting if you're going to live or die. In the world of trauma, it's known as P_s and pronounced *P sub S*. It means the probability of survival. The equation is intimidating:

$$P_s = 1 / (1 + e^{-b})$$

Forget about the math. Here's what you need to know. In an accident, experts say, your probability of survival is a function of your age; your vital signs, like blood pressure and respiratory rate; and the extent of your injuries. The simple fact is that the vast majority of people survive their trips to the ER because their injuries aren't very severe. Of course, others survive against all odds. What makes the difference? Who lives and who dies? I

*For the staggering statistics on how many years of "potential" life are lost every year in the United States, please see Appendix B, "The Arithmetic of Dying Too Soon," on page 345.

ask Dr. Susan Brundage, the trauma surgeon who removed the knitting needle from Ellin Klor's heart and who saved Katharine Decker Johnson after the encounter with the truck. A five foot two blonde from Iowa who wears tortoiseshell glasses and crisp dark suits, Dr. Brundage is blunt, intense, and averse to tooting her own horn. Though it's clear that one critical reason Ellin and Katharine survived is the top-notch medical care that they received, Dr. Brundage says other factors contributed as well. Both women weren't done with life yet and wanted to keep going. They also had very dedicated husbands who took care of them during the toughest times. We'll explore the will to live in chapter 10, but is that the deciding factor? Plenty of people succumb to massive injuries even though they want to live and have loving spouses. Dr. Brundage raises her eyebrows and shakes her head. Some things just defy explanation.

If you're lucky enough to choose which misfortune you encounter, Dr. Brundage tells me, you'd better know the difference between knives, bullets, and brick walls. Let's start with a stab wound to the heart from a sharp object like a knife or a knitting needle. If you show up in the ER with a heart puncture and no vital signs, you have a 37 to 40 percent chance of surviving. Compare that with your probability of survival if you arrive with a gunshot wound to the heart and no vitals. In that case, your chances are just 4 percent. Worst of all, if you've got what's called blunt trauma from smashing into something like a brick wall, your chances are less than 1 percent. The point, Dr. Brundage says, is that every time you get wheeled into the ER, doctors can predict with a high degree of certainty if you'll live or die. And here's what you should remember: Knives are better than guns, which are better than brick walls. A stab in the heart can be repaired with one big stitch; a bullet is more complicated but can still be sewn up. Blunt trauma is the worst, because there's no one fixable thing. It's a whole body of hurt.

* * *

Before Dr. David Spain even begins working on trauma patients, he checks the soles of their feet. It's unconscious, like a reflex, but it's one of his secrets. He sneaks a peek when he approaches a gurney or glances while he's getting briefed on a new case. "If the soles of their feet are white," he says, "we're in trouble." It's usually a sign of internal bleeding and shock, and he knows he's got to do something fast. You met Dr. Spain in chapter 1. He's chief of trauma and critical care surgery and a professor at Stanford. He's also a bullet of a man, pure trajectory, with a shaved head, wire-rimmed glasses, a blue blazer, and a brightly striped shirt. He leads a fast-paced tour of his kingdom, showing me what's called the E2 ICU, a special intensive care unit on the second floor for trauma patients. Near a nursing station, we slow down at a bulletin board covered with a jumble of Polaroid pictures, the success stories of this ward, snapshots taken with patients who come back to thank the trauma team. Dr. Spain shares some of their remarkable stories and then points to the picture of a handsome, tanned man with a mustache who is hugging one of the nurses.

His name is Steve Herrera, a sergeant with the Palo Alto Police Department. A big man, six foot three and 220 pounds, he was riding to work before dawn one morning on his limited-edition Harley-Davidson when an ambulance made an illegal U-turn in front of him. Herrera tried to avoid a collision but ended up sliding headfirst at thirty-five miles per hour into the side of the emergency vehicle. The date was October 2001, and Herrera was forty-four years old. He broke both legs and many other bones, but worst of all he suffered a serious brain injury. For twenty-six days, he lay unconscious in the E2 ICU. Given his age and head trauma, his probability of survival was extremely small. In the waiting room and at his bedside, crowds of friends and cops stood vigil. "I was never alone," Steve tells me. Even when he was "out of it" in his coma, he had a recurring dream.

He could hear someone telling him, "I love you, I'm here for you. I will never leave you." When he awoke three weeks later, he told his wife, Carol, about the voices. "That was me," she said. Those were the very words she whispered every day at his bedside. Herrera is a tough guy—a former leader of the SWAT team who liked to spearhead the charge into dangerous situations—but he tells this story with gentleness and awe. He credits his wife, his children, and his friends with his survival. He also recognizes that the doctors saved his life. "They never gave up on me, either," he says. That's why he returned to the E2 ICU to take Polaroid pictures with the staff and to say thank you.

Six years later, Dr. Spain remembers Herrera, a "big, charming guy" with a severe brain injury. Many thousands of patients later, Dr. Spain even recalls the room number—43—where Herrera lay in a coma. His recovery was "pretty unusual," Dr. Spain says. His chances of survival were in the "single-digit percentages." A scientist and empiricist, he hesitates using the language of the supernatural, but finally admits that Herrera's survival "approaches the miraculous." At age forty-nine Dr. Spain has seen many cases like Herrera's that defy explanation, but he never dwells on them and always moves on to the next crisis.

In the world of trauma surgery, it's the dead who get all the attention. At Stanford and other major centers, doctors compile voluminous reports on those who expire and then carefully review each case. The goal: to analyze if any of the fatalities could have been prevented. In 2006, for instance, there were forty-three "nonpreventable" deaths in the Stanford trauma department. That means a reexamination of all of the facts concluded that nothing could have been done to save the patient. Three deaths were considered "potentially preventable." Under the best circumstances, the patients might have lived. None of the deaths at Stanford was judged to have been "preventable," meaning the patients should have lived. I ask why Dr. Spain and his colleagues don't spend the same amount of time analyzing their success stories like Katharine Johnson and Steve Herrera. He laughs and scratches his bald head. Leaning back in his chair, surrounded by photos of his wife and three sons, he seems a little uncomfortable. He's accustomed to dissecting what goes wrong and probing the unexpected or so-called unpredicted deaths that might have been prevented. No one keeps any records of the most extraordinary cases. He can rattle off all the reasons why people flatline in the ER. But if you press about people who defy the odds and live, he says simply, "That's way outside my comfort zone." It turns out no one ever asks about who survives.

3. The Survival Formula

Every year, a card arrives in Dr. Spain's mailbox. It comes from a military doctor serving at Ramstein Air Base in Landstuhl, Germany. Years ago, the man's wife was undergoing routine surgery when she began to bleed badly. Her physicians couldn't stanch the flow and needed a trauma surgeon to help. Dr. Spain was summoned and quickly stopped the hemorrhaging. Now, every Christmas, a missive arrives from somewhere in the world. Dr. Spain shows me the latest postcard with a beautiful picture of Neuschwanstein Castle, the fanciful nineteenth-century palace in Bavaria that inspired Sleeping Beauty's castle at Disneyland. Scrawled at the bottom of the card are the words the man writes each year: *You saved my family.* Dr. Spain keeps a thick folder of notes and letters like this one, powerful reminders of why he comes to work every day.

After dodging my question about who survives, Dr. Spain admits that he's intrigued about people who live when they're supposed to die, the ones you might call "unpredicted survivors." So after some nudging, he begins to sketch out a survival formula. It's unscientific—guesswork—but it's based upon fifteen years in the trauma trenches. The equation is simple and comes down to three factors in descending order of importance:

FACTOR ONE: YOU

As we learned earlier with P_s, your age and extent of injuries are the decisive factors in trauma cases. Dr. Spain estimates that 85 percent of survival is determined by these two variables. In the math of who lives and who dies, Dr. Spain's survival equation looks like this:

Your Age + Your Injuries = 85 Percent of Your Chances

Younger is almost always better, Dr. Spain says, especially when it comes to head wounds. Indeed, as we saw with Ricky Bunch, the optimum age for a brain injury is between sixteen and eighteen years old. After age twenty, he says, the survival and recovery rates go straight down. Genetics play a role, too. Some people, for instance, are simply more resistant to infections. While some trauma patients succumb to germs after two to three weeks in a hospital, others sail right through without a fever or a problem. In fact, researchers in Newcastle, England, have found that intensive care patients with a particular DNA variant called haplogroup H are more than twice as likely to survive severe infections like pneumonia or the antibiotic-resistant superbug known as MRSA. This variation in DNA—the so-called Survivor Gene—is very common and can be found in around 40 percent of people, especially those of European ancestry.

FACTOR TWO: CIRCUMSTANCE

The second most important variable in survival is situational. Or, as Dr. Spain says, "dumb luck." Where does your accident happen? If your heart attack strikes at Dunkin' Donuts, are the paramedics waiting in line for their morning coffee? When you're crushed by a combine harvester in a cornfield, how far are you from the hospital? When it comes to traumatic injuries, there are three peaks on the graph of when people usually die. The first spike occurs within minutes of the accident. If you're going to perish, it usually happens right at the start. Half of all trauma deaths occur on the spot. Despite all the advances in medicine, those people can't be saved. Indeed, experts say that salvage rates on the scene haven't improved much since the Crimean War in the middle of the nineteenth century. The next peak comes during the so-called Golden Hour, the first sixty minutes after trauma. Although there's no hard proof, paramedics are always taught that the first hour is the most critical. For years, the emergency medicine slogan has been: *Sixty minutes to save a life*. It's not literally true—sometimes you have fewer than sixty and other times you have more. In general, however, 30 percent of fatalities occur when the patient is in shock during the first few hours of trauma. The third spike occurs a few days or weeks later as trauma patients succumb to injuries and infection. In this phase, 20 percent die.

Of course, another situational factor involves the kind of medical care you receive. If a truck flattens you, who puts you back together in the ER? Are you lucky enough to have a surgeon like Dr. Susan Brundage who trained at the University of Washington in Seattle, a major trauma center that was responsible for the worst injuries from five surrounding states? Dr. Brundage spent years handling complex cases, so she knew exactly what to do when Katharine Decker Johnson arrived. It may seem obvious, but not all doctors and emergency departments are created equal. Trauma doctors use something called the Injury Severity Scale (ISS) to track a patient's multiple wounds. The math goes from 0 to 75, and the higher the number, the less survivable your injuries. If your score is above 25, for instance, your chances of surviving in a typical ER are between 70 and 75 percent. At Stanford, your chances are 85 percent. When it's your neck on the line, that's a big difference, and it's why Dr. Spain believes circumstance can save your life or kill you.

FACTOR THREE: INTANGIBLES

Dr. Spain admits that survival isn't always determined by physiology and situational factors. He's seen many cases influenced by X-factors that just can't be measured. You might call these the mysteries of the emergency room, and doctors usually don't pay much attention to them. "If we can't define it," he says, "we tend to ignore it." For instance, personality plays a small but important role. A "rotten old bastard that's tough as nails" often fares better than a sniveling wimp, Dr. Spain says: "They fight to the end." As a scientist, he knows this generalization is pretty flimsy. It's only based on anecdotal evidence, but he's seen enough cases to believe that personality plays some kind of role. To bolster his opinion, he cites a Dutch study showing that "dispositional optimism"—a positive mental attitude—improves survival rates in older patients. Optimistic people live longer than pessimistic people, the study found, and the "protective" health effect is even more pronounced in men than women.

Another X-factor, as we saw with Ricky Bunch and Steve Herrera, is the support of friends and family. Again, it's unquantifiable, but Dr. Spain believes there's some kind of correlation between the size of the crowd in the waiting room and the chances that a patient recovers. It's definitely not a guarantee. Plenty of people have perished surrounded by a horde of loved ones. But again, there's something powerful about family and friendship. He's seen the effects too often to discount the impact.

The final X-factor in his survival equation is faith. Catholic by upbringing but lapsed in his observance, Dr. Spain believes some people seem to be protected by a higher power. Again, it's beyond the realm of science and empirical proof, but a framed newspaper clipping from *The Courier-Journal* in Louisville, Kentucky, suggests that he's a believer. It hangs next to his desk and tells the extraordinary story of one of his patients. "I swear to God that guy had an angel watching over him," Dr. Spain says.

Around ten in the morning on February 25, 2000, Gary McCane Jr. was working on the air-conditioning and heating of a new home in Woodmont, Kentucky, a subdivision of Louisville. Someone came running to say that two workers were trapped nearby in an underground cistern. A self-described "skinny guy," Gary volunteered to shimmy down the hole and rescue the men. As soon as he started down the ladder, a huge explosion rocked the subterranean tank. "I can actually remember being in the fire," Gary tells me in a slow drawl. He managed to scramble out of the hole, roll on the ground to extinguish the flames, walk across the street, and sit down on the curb. His cotton T-shirt, green work pants, and one boot were incinerated. The only clothing left on his body was his leather belt, the waistband of his pants, and his other boot. Waiting for the ambulance, he felt incredibly cold and desperately wanted a blanket.

The cistern explosion—caused by an electrical spark—killed the two other workers, and Gary suffered "full thickness" or third-degree burns on 85 percent of his body. Even though he was twenty-one years old and in excellent physical condition, doctors believed his prognosis was uncertain. He spent the next two and a half months in a medically induced coma, undergoing forty surgeries at University Hospital in Louisville. He still remembers a vivid dream from those unconscious days. He saw himself walking down a sterile white hallway, passing a nursing station, and arriving at the exit. In his dream, he tried to open the double doors but they wouldn't budge, so he returned to his room. Later, he found out that his grandfather—known as Pappaw—had gathered a group of fellow Pentecostal ministers to pray in the waiting room. On the night that Gary "coded"—hospital-speak for cardiac arrest—Pappaw and his friends had laid hands on the exit of the hospital. They asked God to help Gary heal before he left the burn unit. In a coma, of course, he had no way of knowing the layout of the hospital, including the location of the nursing station, hallway, and doors. And yet his dream was accurate in every detail. Today he wonders what he would have seen beyond those exits that wouldn't budge. Was death waiting for him there? Did his grandfather's prayer keep those doors firmly shut? "I don't know," he says.

Despite the severity of his injuries, Gary's recovery was so quick that he was released from the hospital one year before the doctors had predicted. For his bravery in trying to save those two other workers, he was honored with a medal and thirty-five hundred dollars from the Carnegie Hero Fund Commission. "I'm not sure I deserve it," he says now. He doesn't believe he did anything special except for getting blown up.

4. The World's Most Exclusive Survivors Club

To test the Survival Formula, let's consider a rather extreme scenario. Imagine if you suddenly took leave of your senses and decided to jump off the Golden Gate Bridge. From the moment you leaped, you'd have four seconds before hitting the green water 240 feet below. You'd reach a top speed of seventy-five miles per hour before slamming into the Golden Gate Strait. On impact, you'd feel a bone-crushing force of fifteen thousand pounds per square inch that would rip your internal organs loose. Depending on your angle of entry, your ribs would splinter, spearing your spleen, lungs, liver, and heart. Your head would fracture, your neck would break, and your pelvis would snap. When you were pulled out of the water, you would probably look okay on the outside, but inside it would be a whole different story. Ken Holmes, the Marin County coroner who handles Golden Gate suicides, says that when you lift an average corpse it's stiff like a board because of the bones and muscles holding everything in place. The bodies of bridge jumpers, by comparison, feel like big sacks of BBs or rice. They're soft and limp because their interior scaffolding has been shattered.

One myth about jumping off the bridge is that it's a beautiful, peaceful, and guaranteed way to go, Holmes continues. The 1.7-mile orange span is one of the Seven Wonders of the Modern World, and when you stand on the edge, you look out at the San Francisco Bay and Alcatraz Island. "In reality," Holmes says, "you're ripping your body asunder from the inside out. It's not pretty, it's not painless, there's nothing nice about anything that happens once you step off that bridge." Often, a jumper's heart is ripped from the aorta. If they don't splinter, your twelve ribs in front and twelve in back compress like a bread slicer, dicing your organs. Even if you're still alive after the initial impact, you're almost certain to drown. The strait is 350 feet deep, and jumpers can plunge some 80 feet below the surface. If you even make it back up for air and your lungs haven't been shredded, the currents are so powerful they'll sweep you away. The bodies of jumpers often wash up many miles from the bridge.

It's a miserable distinction, but the bridge is reportedly the world's number one suicide destination. Some mental health experts even call it "a loaded gun in the middle of the city." Once every couple of weeks—or fifteen days on average—someone hops over the rails or leaps off the chord, the thirty-two-inch outer beam of the bridge. Since it opened in May 1937, at least 1,250 people have taken their lives this way. The typical suicide is 41.7 years old. The youngest was a fourteen-year-old girl and the oldest an eighty-four-year old man. Ninety-eight percent of the time, according to the *San Francisco Chronicle*, the jumper dies. Compare that with the fatality rate of poison, which is 15 percent; a drug overdose, 12 percent; and wrist slashing, 5 percent. Only 28 jumpers are known to have survived the fall. It may be the world's most exclusive Survivors Club.

Kevin Hines's interior monologue went something like this: *I'm a terrible man. I have to die. I have to leave my family alone because I'm going to hurt them.* When he dove off the bridge on September 24, 2000, he hadn't really slept in two weeks. More than one year earlier, he had been diagnosed with bipolar disorder, frequently called manic depression, and he struggled with the cycles of up and down. Life was a roller coaster of mania then depression, mania then depression, paranoia then mania then depression. Hines couldn't stand looking in the mirror. The nineteen-year-old hated his reflection, and he began to obsess about suicide.

Hines wrote six or eight suicide letters in a notebook before settling on the right words. He didn't want to be too mean to his parents, who had adopted him as an infant. He addressed his farewell letter to them along with his sister, brother, and best buddy. That morning his father, Pat, dropped him off at City College in Ingleside, a San Francisco neighborhood. Hines kissed him on the cheek, a family tradition, and said, "I love you." His dad answered: "I love you, too, Kev. Be careful. Have a good day." Hines knew it was the last time they would see each other. He stopped by his English class, where he signed his suicide note: *Sincerely, John Kevin Hines. Please forgive me.* Then he got up from class, said good-bye, took a bus to Walgreens, where he stole his last meal—Skittles and Starbursts—and boarded another bus for the short ride to the bridge.

Hines says the voices in his head "started banging really loud": *You have to die! Jump now! You're not a good person!* Then a more "rational" voice would take over, saying: *I don't want to die. Get off the bus now.* The first voice would fire back: *Shut up! You're going there and you're going to die.* Hines wept quietly in his seat in the very back. He hoped someone would ask: *Are you okay? Is something wrong? Can I help you?* If anyone had shown even the smallest sign of caring, he would have told them everything. If a single human being had reached out that day, he wouldn't have gone through with the jump. Hines says these thoughts are common to suicidal people, but no one bothered to extend a hand or utter a word. When the bus reached its final destination on the bridge, everyone got off, but Hines didn't move for a few moments. He was still crying and hoping the driver would just ask if he was all right. Instead, the man at the wheel shouted: "*Get off.*" Bawling, Hines wandered around the bridge for forty minutes, desperate for someone to say anything. Again, no one even noticed.

At 9:45 AM, he finally found "the spot" to jump near the 101st piling of the bridge. The voices in his head were debating furiously: *Jump, stop, jump, stop!* Suddenly a beautiful woman in big fashionable sunglasses approached him. Hines thought: *Here's my savior. This is it. Fantastic!* In a European accent, the woman asked: "Will you take my picture?" Hines was shocked. He thought: *Lady, can't you read my mind? Don't you know what's going on? Can't you see the tears flying down my face?* Hines snapped a few photos and exchanged pleasantries. *Nobody cares,* he told himself. Then he heard the voices say: *Jump now! Do it! Go!* "I ran and catapulted over," he recalls. "I just put my hands on the rail and threw my entire body over."

As Hines plunged headfirst toward the water, his very first thought was: *I don't want to die!* Raised Catholic, he hadn't gone to church in years. Still, he called out to God. "I felt bad that I was jumping on the God bandwagon right when I needed His help," he remembers. His second thought: *I'll never survive if I hit the water headfirst. If I don't really want to die, I better get my feet first and see what happens.* So he tried to throw his head back, which shifted his weight and turned him around. It may have been this effort or a gust of wind, as doctors speculated later, but somehow he hit the strait feetfirst at a forty-five-degree angle in a sitting position.

The impact was excruciating. Two vertebrae—his T12 and L1—popped instantly, splintering and lacerating his lower organs. Doctors told him later that if he had fallen a smidgeon differently, he would have severed his spinal cord. He had always believed that if you jump off the bridge, "You just kind of die. You just kind of vanish into thin air." But it wasn't true. Hines found himself forty to fifty feet underwater with his lungs burning for air. He didn't know what to do or which way to swim. His legs wouldn't move, and his arms had been wrenched violently on impact. "I think I went right," he remembers. "Then left. Then down. I turned around in the water and finally I saw the light." With his head throbbing, he broke the surface just when he thought he would pass out.

Six powerful currents collide and roil under the bridge, and Hines was dragged into the vortex. *Now I want to stay alive,* he thought. *I have to live . . . but how am I going to live? My legs aren't working. And I can't move.* Unable to stay afloat, he thought: *This is it. I'm going to die here. I'm going to die in the water. I don't want to die drowning. This is terrible. I can't drown. I will not drown.* Hines looked up and was stunned by the size and scale of the bridge. "When you're looking down at the water," he explains, "it's kind of like looking down at your carpet in your house. It doesn't look that far away because it's such a big body of water. When you're looking up from the water to the bridge, it's one of the scariest things you'll ever imagine if you're in a bad way." So he started praying: *God, please save me. I don't want to die. I'm so sorry for what I did. Please save me.*

Incredibly, he felt something bump into him. *Oh man, oh man. It's a shark,* he thought. *This thing is just going to devour me. It's going to bite my legs.* He tried punching it to make it go away, but the creature kept nudging and circling him. "It kept going around me and bumping into me and going around me," he remembers. "And at the moment, I lay afloat on top of the water. I wasn't moving my arms. I wasn't moving my legs. I was just lying there, and this thing was brushing by me and swimming underneath me." Hines found out later it was a sea lion that may have sensed he was in distress and tried to help. "A friend of mine in the Coast Guard thinks it did keep me afloat," Hines explains.

When a Coast Guard ship approached with loud engines, the sea lion took off. Hines feared he would be run over or cut in half by the propeller. "They knew exactly where I was," he says. "I think two gentlemen jumped in the water and kind of hoisted me up." An ambulance rushed him to Marin General Hospital. As one of the paramedics bid farewell, he said: "Good luck, kid." His internal injuries were massive, and doctors told him he had a fifty–fifty chance of surviving the night. But Hines thought: *Screw that! I'm going to fight. I'm going to live through the night. Not only am I going to live through the night, I'm going to live to be 110. Nothing like this is ever going to happen again.*

Today Hines is married and works in San Francisco as an administrative assistant at the Alliant International School of Psychology. He still suffers suicidal thoughts every now and then, but he's a man on a mission. He has joined the National Mental Health Awareness Campaign to teach high school and college students how to take care of their psychological problems. He is also a director of the Bridge Rail Foundation, which campaigns to install a suicide prevention barrier on the bridge. His last stay in a mental hospital was in 2004 and he's learned a lot about surviving. "There are 101 things I need to do every day to stay focused, stay healthy, and to stay above water per se," he explains. "It started with being disciplined in brushing my teeth every day and taking a shower every day. I'll just leave it at that . . . I am forward thinking right now. I stay in the present. I know what I have to do today, but I'm hoping for tomorrow."

Who survives jumping from the Golden Gate Bridge? The answer is simple: Young men with good muscle tone who strike the water feetfirst at a slight angle, says Ken Holmes, the Marin coroner who has worked with the dead for more than thirty years. Since it's the equivalent of leaping from a twenty-five-story building, he explains, your entry into the water makes all the difference. If you hit at a small angle, your feet, ankles, and knees bend to absorb some of the impact and your body arcs through the water without going too deep. If you strike feetfirst at too steep an angle, your legs will take the brunt of impact, but you'll probably go straight down and drown. If you hit headfirst, you're dead because your cranium takes all the blunt force. Same with landing on your belly, back, or side. At seventy-five miles per hour the sudden stop finishes you off.

So how did Kevin Hines survive? If we consider the first factor in Dr. Spain's Survival Formula—age and injuries—we know that Hines was young and physically fit. Instinctively, he knew he would die if he hit the water headfirst. Within seconds, he took action to shift positions. Perhaps that decision reduced the extent of the trauma and saved his life. The second factor in the Survival Formula—circumstance—also played a role. That gust of wind may have helped him flip into the feetfirst position. And the sea lion may have kept him afloat until the Coast Guard arrived. Finally, the third dimension of the Survival Formula—X-factors—may explain why he lived. Hines believes deeply that the Almighty delivered him from oblivion. God gave him a second chance, he says, sending an angel in the form of a sea lion to rescue him.

Clearly, the Survival Formula is dynamic. At one level, living and dying are up to you, and there are practical steps you can take to improve your chances. Do you take reasonable care of yourself, storing up physical and emotional reserves to depend on in a crisis? Do you know where the nearest ER is located? Do you know first aid and CPR? Do you maintain relationships with family and friends who could rally around you in a time of need? At another level, admission to the Survivors Club depends on sheer circumstance, twists of fate, and the randomness of life. In the chapters that follow, I'll explore some of these other dimensions. For instance, does prayer actually make a difference in who lives and who dies? Do certain people really have *all* the luck while others don't? In short, how much of life—and death—do you really control?

Survival Secrets

The Best Place to Suffer a Heart Attack

Every twenty seconds, a heart attack strikes someone in America, killing five hundred thousand a year. That's fifty-seven deaths every hour, almost one per minute. In the United States and many nations, it's the leading cause of death among adults over age forty. The worst kind of heart attack—called sudden cardiac arrest—occurs when your ticker's electrical system goes haywire, and stops pumping blood to the rest of your body.* In major cities like Los Angeles, Chicago, and New York, your chances of surviving cardiac arrest are less than 3 percent. In communities with the best emergency response systems like Seattle and Boston, the salvage rate tops out at 9 percent.

So who survives? People like Thurman Austin. The sixty-three-year-old textile worker from China Grove, North Carolina, was gambling at the Stardust Casino in Las Vegas when he collapsed, banged his head on a dollar slot machine, and hit the floor. He didn't even know his wife, Gwen, in the next seat had just won nearly three hundred dollars. Within minutes, security guards arrived with one of the casino's new defibrillators and shocked his heart back into normal rhythm. It was July 1, 1997. "I hit the jackpot that day," he says.

Yes, he did. Of all the places in the world—believe it or not—casinos are the safest place for a heart attack, according to the *New England Journal of Medicine.* The reason has nothing to do with Lady Luck, says Dr. Bryan Bledsoe, a former paramedic and emergency physician who teaches at the University of Nevada–Las Vegas and writes textbooks for emergency caregivers. Survival depends on how fast you're defibrillated and receive chest compressions. If you get the first jolt within one minute, your chances are around 90 percent, but they drop 10 percent every sixty seconds. Incredibly, the salvage rate on the

---

* I use the term *heart attack* in its most common sense. Technically, a heart attack is a myocardial infarction or blockage in your heart's plumbing while cardiac arrest is an electrical problem with your heartbeat, according to the Heart Rhythm Association.

Vegas Strip is now 53 percent. Even in a hospital, your odds aren't as good.*

Why is Sin City the best place to survive cardiac arrest? It starts with the alarming fact that two to three times more people suffer cardiac arrest in Las Vegas than other cities of similar size. The reason: Vegas Syndrome. Older tourists keel over because of too much eating, drinking, partying, smoking, exhaustion, and stress from gambling. Indeed, paramedics in Clark County, Nevada, found themselves responding to so many fatal cardiac arrest calls in casinos that they urgently needed a solution. In 1997, they began persuading casino owners to buy automated external defibrillators and to install them like fire extinguishers in public places. They also started training casino workers in CPR and defibrillation.

With security cameras and guards always on the lookout for cheaters and troublemakers, virtually everyone is under constant surveillance. That means if a visitor drops, someone notices quickly. If you keel over at the MGM Mirage, for instance, a trained staffer with a defibrillator will be standing over you in just 2.8 minutes. Even if you're in a hospital, the response rate isn't always so fast. And that can make the difference between hitting the jackpot and losing everything.

---

* In 30.1 percent of sudden cardiac arrest cases, deadly delays occur in defibrillation because the medical staff doesn't move fast enough, according to a recent study. And beware of nights and weekends, which are substantially more lethal than weekdays for patients with in-hospital cardiac arrest, according to another new study.

5

The Supersonic Man

HOW MUCH OF LIFE (AND DEATH) DO YOU REALLY CONTROL?

As you sit reading this book, you may not realize that you're experiencing 1 g of acceleration, the technical way of saying you're subject to the normal force of gravity. If you move slightly, any change in your speed or velocity is measurable in units of acceleration known as g. For instance, when you cough, you experience 3.5 g, if only for a split second. The intensity or impact of acceleration on your body is known as g-force. If you weigh 170 pounds, coughing translates into a very brief g-force of 595 pounds. Everything you do in life generates some kind of acceleration or g. A sneeze produces 2.9 g and a slap on the back, 4.1 g. These bursts last such a short time that they're barely noticeable.* That's why you don't hurt yourself when you plop into a chair—10.1 g—or hop off a step—8.1 g. The more g you experience or "pull," as aviators say, the greater your

* During a sneeze, you expel air (and germs) at more than a hundred miles per hour. Sternutation, as it's known medically, can be dangerous. In July 2006, Dean Rice, an eighteen-year-old Welsh teen, was camping with friends and family when he experienced a sneezing fit. He brushed it off as allergies, but within minutes he had fallen to the ground and died of a brain hemorrhage caused by sneezing. In May 2004, Chicago Cubs slugger Sammy Sosa sneezed twice so violently that he sprained a ligament in his lower back and ended up on the disabled list for fifteen days.

chances of injury. A body can handle 18 sustained g before the lungs compress, breathing becomes difficult, and internal organs begin to rip apart.

Most of us are rarely subjected to more than 8 g for a second, let alone two. But if you're a military pilot, g-forces are part of the job description. They're also a serious hazard. Inexperienced aviators can black out between 4 and 6 sustained g. The best fighter pilots can handle 9 g before they experience what's called g-LOC, gravity-induced loss of consciousness. A typical 195-pound flier pulling 9 g feels like he weighs 1,755 pounds. As you can imagine, that makes controlling a plane very challenging: Your arms feel like barbells, your head turns into a concrete block, and *brain drain* takes on a whole new meaning. Under extreme downward g-forces, the blood rushes from your head toward your legs, but your heart isn't strong enough to pump it back to your brain. Within seconds, you lose the ability to see color, a condition known as brown-out or gray-out. Next come tunnel vision, blackout, and a trip to a place that pilots call Dreamland.

The military lingo for ejection is quite evocative: It's called riding the rocket because you're literally shot from the plane by a series of meticulously synchronized explosions. As soon as you pull the ejection handle, one ballistic charge catapults you from the cockpit, another rocket blasts you to a safe height in the air, and then you're left dangling beneath a parachute that opens automatically from another charge. It's an incredibly violent experience. At typical altitudes and speeds, after ejecting you slam into a wall of wind with a force of more than three tons. The aerodynamic pressure or stress is so intense that some fliers claim you can permanently shrink an inch or two from spinal compression. Perhaps the greatest danger comes from "flail," when the windblast can dislocate or break your limbs, spinning them like pinwheels.

At the Navy's Survival Training Center in Miramar, there's a tower next to the pool where they introduce sailors to g-forces and ejection seats. Inside the tall shed, you'll find Device 9E6, the Universal Ejection Seat Trainer. It's a big fifty-five-hundred-pound contraption that looks like some kind of ancient siege engine or catapult. It's got a wide base with thirty-foot metal rails rising at a steep seventy-two-degree angle. At the bottom, an ejection seat is attached to the track. Driven by pneumatic pumps, a rider shoots backward up the slide, experiencing accelerations up to 6 g, less than half the speed of a real ejection.

I'm usually okay with roller coasters, but as they buckle me into the ejection seat, one word from our classroom briefing sticks in my mind: *jolt*. That's what they call the initial shock when the shot begins. On this machine, it's supposed to register around 60 g. In a real ejection, it's around 200 g. It lasts only a millisecond, but I wonder if my body can take it. What if something important shakes loose? My heart thuds a little faster and a few drips of sweat trickle down my face. It's very warm in all this equipment. I'm wearing a helmet, breathing mask, flight suit, parachute harness, survival vest, gloves, and anti-g pants. The pneumatic pumps start to whir, and they lower my seat into the launch position. It feels like I'm in the pouch of a giant slingshot. The instructor tells me to press my head back into the seat, elevate my chin ten degrees so that my neck doesn't snap forward, and stare straight at the red cross made from duct tape stuck to the wall in front of me. I'm distracted for a moment by the symbol of mercy—are they sending a message?—then the trainer reminds me to keep my elbows and legs tight during the ride. On command—"Eject! Eject! Eject!"—I take a deep breath and pull hard on the yellow-and-black-striped handle.

Wham.

I notice the sound but barely feel a thing. I'm jerked six feet up the rails, and then it's over. Just like that. In a real ejection, I'd already be floating beneath my parachute. But now, as they lower me back down to the starting point, I feel a little embarrassed. *That wasn't so hard. Why did I break a sweat?* The instructor asks urgently: "Are you free from all pain and discomfort?" I nod, give the *Top Gun* thumbs-up sign, and they unhook my harness and lift me from the seat. I fumble trying to remove my mask and helmet and notice that the trainers have gathered in a huddle. Finally, one of them comes over to inform me that my leg flailed going up the rails and my steel-toed boot tripped an alarm. In a real ejection, I would have struck the plane on the way out, resulting in a serious injury or worse. They blame it on my size. Someone mumbles that I'm too tall for the ejection simulator Then they hand me a computer printout with a graph of the g-forces I experienced during my ride. Sure enough, the initial jolt was 57 g, but it only lasted a snap. The average or sustained g load during the entire ride was 3.5. I stare in disbelief. The force of my shot was only as powerful as a cough and barely stronger than a sneeze. It's more humbling than deflating, especially given the story they like to tell around here about the Supersonic Survivor.

1. "Bail Out!"

The second rule of the Survivors Club tells us that it's pointless, even foolish, to make comparisons of adversity. At risk of breaking my own rule, I believe one story stands out for its sheer intensity and brutality. Of all the survivors I interviewed, Brian Udell probably suffered the single most sudden and violent ordeal. Udell is the only pilot ever to survive ejecting at sea level from a jet going faster than Mach 1, the speed of sound.* Incred-

* A handful of aviators have survived supersonic ejections at higher altitudes, like SR-71 reconnaissance pilots who have bailed out flying above seventy-five thousand feet where the air is so thin that the impact forces are actually less than those encountered ejecting at five hundred miles per hour at sea level.

ibly, he endured a sustained load of 45 g. Given his weight—195 pounds—that means he faced g-forces of nearly 9,000 pounds, the equivalent of an RV trailer parked right on top of him. His tale reveals the remarkable strength of the human body. It also underscores how much we can control in survival situations and how much we can't.

I meet Udell in the Sheraton Hotel lobby near Los Angeles International Airport. He's a captain with Southwest Airlines, and he's on a short layover. He wears a polo shirt, khaki shorts, and running shoes. He's six foot one, very fit, and his dark hair is cut short, parted on the side, and gray at the temples. Despite an Arizona tan and easy smile, he's got a precise and upright bearing that hints at his decade in uniform. As we sit down in the lounge area, he crosses a leg and points to a bump on his knee. On closer inspection, it's actually a screw poking beneath the skin. Udell has plenty of scars from his ordeal: His left ribs still stick out where they shattered, and a muscle in his chest is still bunched up in a lump where it tore. He's got six screws in his right knee and six more plus a metal plate in his left ankle. It may sound like a lot of retained hardware—the air force's phrase for the nuts and bolts they leave inside after they fix you up—but Udell knows it could have been a lot worse.

On April 18, 1995, he was flying an F-15E tactical jet fighter off the coast of North Carolina. It was a routine training exercise on a moonless night. A captain in the air force, Udell was known by the call sign Noodle, a twist on his last name Udell. An experienced pilot who has flown more than one hundred combat missions, Udell also served as an F-15E instructor, teaching new aviators to operate the forty-seven-million-dollar airplanes. On this training run, Udell's jet and another plane were playing the aggressors—Red Air as they're called—and were following a carefully choreographed Soviet-era combat strategy to attack two other F-15s. Udell's flight call sign was Sword 93, and his role was to serve as the decoy—or "duck"—in this aerial encounter. As planned, he began an easy maneuver—a sixty-degree right turn—to distract his adversaries while his wingman snuck in for the kill. Udell checked his heads-up display, a transparent piece of glass like an instrument panel that enables a pilot to check crucial flight information while simultaneously looking outside. The display indicated he was flying 460 miles per hour at twenty-four thousand feet and banking right at sixty degrees.

Udell sensed something wasn't right when he heard wind rushing over his canopy, the Plexiglas bubble covering the cockpit. Usually that whooshing sound meant the F-15 was traveling around six hundred miles per hour, much faster than the display indicated. Since learning to fly at age nine in Texas, Udell was trained to trust his instruments but also to listen to his airplane. His father was a celebrated air force colonel who had won the prestigious William Tell competition, a biennial top gun contest using real ammunition. At age ten, even though he wasn't tall enough to see out of the cockpit for takeoffs and landings, he helped his father pilot a Cessna 190 across country from Houston to North Carolina. From those early days, he knew that his planes talked to him in ways the gauges and dials sometimes couldn't. On this night, he was sure the computer was wrong.

Udell checked another set of displays near his knees. He needed to know which way his plane was pointing. Staring at the attitude direction indicator, he realized his airspeed was now 690 miles per hour and his altitude was seventeen thousand feet. In a matter of just seconds, he had accelerated 230 miles per hour and lost seven thousand feet of altitude. *Holy smokes,* he thought, *I'm pointing straight down at the earth.* To help recover a plane in a difficult situation, pilots are trained to hand over the controls to the weapon systems officer—or wizzo—who also acts as navigator. Dennis White was sitting in the backseat that night. He had been Udell's partner for a couple of years, and they had flown a handful of missions together in Iraq. Even on a routine training run, their lives were literally in each other's hands.

"What's going on?" Udell asked, hoping Dennis would have a fix on the problem.

"I don't know," he replied. "But we need to recover."

"What is our attitude?" Udell said, using pilot lingo for the direction they were pointing.

"I don't know."

I'm on my own, Udell thought. Dennis wouldn't be able to recover the plane. If the F-15 was really flying straight down toward the ocean, he only had a few more seconds to solve the problem or it would be too late. *Oh my gosh,* he thought, *this is really happening.* The entire drama—from that simple right turn to a life-or-death situation—had taken only five or ten seconds, less time than you need to read this sentence. Udell buried his face in his instruments one last time. The jet was upside down, slicing toward earth at an almost vertical angle of eighty degrees. At 10,000 feet, Sword 93 shattered the Mach 1 barrier of 1,116 feet per second or 769 miles per hour. Udell realized it was too late to save the plane. Now it was time to save their lives. *We've got to go,* he told himself. *We've got to get out of here.*

"Bail out! Bail out! Bail out!" Udell commanded. In the time it took to pull the ejection handles, Sword 93 traveled another four thousand feet straight toward the ocean. The plane was now just six thousand feet above the water and would plunge into the Atlantic in less than six seconds. Udell pressed his head back against the seat and tucked his elbows in as tight as possible. The next sensation was surreal. He watched the canopy slide back. He saw a white flash of light and an enormous wind blast. And then there was only darkness.

A ground radar station in North Carolina sweeping the sky captured the unbelievable facts of Udell's ejection. With Sword

93 slashing straight down toward the ocean, the bubble covering the cockpit ripped away at forty-five hundred feet. Just a second later, as the plane plummeted through three thousand feet, Dennis White ejected. At fifteen hundred feet, Udell got out. Radar hits show that his parachute opened just five hundred feet over the water. In the darkness of the Atlantic, Udell could hear the flutter but couldn't see a thing. Above and below, the night was black, and he had no way of calculating how soon he would plunge into the ocean. His body ached all over—a dull pain—and he felt exhausted. Until that point in his life, his worst injury had been a sprained ankle. Now, moments before ditching in the water, his body was limp as if it had been "hit by a freight train." But his training and instincts kicked into gear. *Better get my stuff together real quick,* he remembers thinking. The cool night air felt so strange on his face. Then he realized his helmet and mask had been ripped off by the windblast. In the hospital, he would learn that all of the blood vessels in his face had exploded, his lips swelled up like hot dogs, and his head inflated to the size of a watermelon. Falling toward the water, Udell rushed through the post-ejection checklist ingrained in his mind. The life preserver around his neck was of no use—it had been sliced into ribbons during the ejection. His gloves and watch were gone, too, ripped off by the force of the ejection. A one-man life raft was supposed to be hanging at the end of a fifteen-foot cord attached to his right hip, and he prayed that it hadn't been shredded. He tried to move his left arm, but it wouldn't budge. Badly dislocated, it was pointing in the wrong direction, bending outward ninety degrees at the elbow like the little hand on a clock pointing toward 3. With his right hand, he jerked the line of the life raft to make sure it had inflated. Pulling up with his one functioning arm and gripping stretches of cord in his teeth, he hoisted the raft as close as possible. With all of his injuries, this was his only hope when he hit the frigid water of the Atlantic.

* * *

One moment he was dry. The next, he was ten feet under. Udell felt the salt water burn his wounds, and he struggled to the surface. Immediately the wind grabbed his parachute and began to drag him across the ocean. Within seconds, automatic explosions blew away the lines connecting him to the chute. Now he was alone some sixty-five miles off the North Carolina coast in five-foot seas without a life vest when he noticed "green sparkly stuff" all around him, like fairy dust. At first, he was frightened, wondering if he was seeing this eerie glow because of a brain injury. Then—incredibly—he remembered a *Discovery* program on TV about ocean phosphorescence and plankton that gives off light when it's disturbed. "It was the coolest thing," he remembers. For a second, he actually appreciated the extreme beauty of the night. Then he told himself, *Okay, I've got to get back to business.*

First, he tried a frog kick and realized how badly his legs were damaged. His thighs moved on command, but below the knees, everything flopped in the wrong direction. His right knee was horribly dislocated. All the tendons and ligaments were severed; only the artery, vein, and skin were holding his shin and foot to his body. His left leg was a mess, broken below the knee, and his whole foot pointed backward. Three of his four limbs didn't work, and he quickly realized that swimming wasn't really an option. He tried to pull himself onto the life raft, but with only one functioning arm, he couldn't get leverage. Every time he hauled himself up onto the lip of the raft, a wave knocked him off. He compares it to the awkward struggle of climbing onto an inflatable float in a swimming pool, except he was battling with one arm in high seas.

Udell knew he was burning through adrenaline and wouldn't be able to keep going for much longer. Panic began to set in. Sharks couldn't be too far away, and they surely smelled blood in the water. Hypothermia was another threat if he didn't drown first. He tried again and again to climb onto the raft but the seas were too rough. Finally, he put his head against the canvas, closed his eyes, and said to himself: *This is it. I'm going to die tonight.* His eyes well up with tears as he remembers his decision to stop fighting for his life and to start praying. Military training had carried him this far, but now he put his fate in the hands of a higher power. This was the "defining moment of the whole experience," he says. Broken and battered, he cried out: "God, I need help."

2. The Doctor of Extremes

Brian Udell's story, which we'll return to shortly, presents a critical question: How much punishment can the human body really tolerate? In other words, what are the outer limits or farthest boundaries of survival? In search of answers, I paid a visit to a Gothic brick castle on East 70th Street in New York City, not far from Central Park. The Explorers Club is one of the most unusual and prestigious organizations in the world, a rarefied place where the members know all about going to extremes and surviving against the odds. A plaque at the entrance announces some of their historic achievements: Robert Peary, first to the North Pole in 1909; Roald Amundsen, first to the South Pole in 1911; Sir Edmund Hillary, first to conquer Mount Everest in 1953; and Neil Armstrong, first to set foot on the moon in 1969. Founded in 1904, the club boasts some three thousand members around the world. Each is a leader in exploration and discovery.

In the wood-paneled library where six-foot elephant tusks curve like parentheses at the ends of a fireplace, I meet Dr. Ken Kamler, the club's vice president for education and scientific research. In his day job, Kamler is a microsurgeon in Long Island, New York. He fixes hands and fingers, relieves carpal tunnel syndrome, repairs nerves and tendons, and sets fractures. But his real passion—and more exotic specialty—is what you might call extreme medicine. At age sixty-one Kamler is one of the world's preeminent expedition doctors, traveling with teams of adventurers to some of the most distant and dangerous places on earth. If you get frostbite at the South Pole or crack your head in the Amazon, he knows what to do. He can treat crocodile bites and five-hundred-volt electric eel shocks. He can handle venomous iridescent caterpillars or tunga fleas that burrow under your skin—especially your bottom—and itch like crazy. He knows what to do if you encounter a poison-dart frog—a phyllobate—that produces some of the most deadly natural toxins in the world. And at high altitude, he's got the right ointments for sunburn in the unlikeliest spots, like inside your nose and under your eyelids. In May 1996, he was preparing to summit Mount Everest when a devastating storm overwhelmed three teams of climbers coming down from the peak. Twelve mountaineers perished, and Kamler treated the survivors as they stumbled into his camp on Everest's icy Lhotse Face. For his heroism and altruism at twenty-four thousand feet, he received an Explorers Club award. "My patients are people who live on the edge of survival and beyond," he writes in his excellent book *Surviving the Extremes*. "I practice medicine in environments and in situations incompatible with life, often treating conditions I have never seen before, and sometimes never even imagined."

After meeting Brian Udell and many others who survived the impossible, I've come to ask Dr. Kamler two questions: How much can the human body endure? And how much of life and death do we really control? Kamler seizes upon each question with energy and enthusiasm, and his answers are both practical *and* paradoxical. Our bodies are both strong *and* vulnerable, he explains. Our physical limitations mean we can only live on one-fifth of earth's surface, where temperatures range between zero and one hundred degrees Fahrenheit. Above an altitude of eighteen thousand feet, we can't survive on a permanent basis.* "Beyond that," he explains, "the environment is too extreme for an organism that needs food and water daily, oxygen by the minute, and heat constantly."

Twirling a pair of eyeglasses, Kamler explains that five out of six people on earth depend on civilization to stay alive. Without eye doctors and corrective lenses, many of us wouldn't be able to see, let alone earn a living. Without supermarkets, we wouldn't be able to feed ourselves. Without sewage systems and water treatment facilities, many of us would succumb to cholera and other diseases. "All of us are descendants of survivors," he says. "Otherwise, we wouldn't be here." But with modern society protecting and sustaining most of us, he continues, "we don't really know if we have that instinct for survival anymore." Kamler believes in an interesting duality. On the one hand, we're "far more fragile than we'd like to admit. If our protection breaks down, we die easily." On the other hand, he says, he's witnessed the "the body's enormous capacity for survival."

Every once in a while, Kamler sees a survivor who defies the laws of medicine and physics. Some people call these instances miracles or supernatural events. He thinks of them as natural occurrences that simply go beyond our current ability to comprehend. One of them occurred on his first expedition to Mount Everest in 1992 when a Sherpa named Pasang fell headfirst eighty feet into a crevasse. In his late twenties, Pasang was one of the Nepalese porters carrying equipment and supplies for a team of climbers from New Zealand. He was working without a safety line and landed upside down with his head wedged into the bottom of the chasm. Rescuers managed to tie a rope around his waist and pull him up. Incredibly, he seemed okay. His nose was bleeding, but he was able to walk. Then suddenly

* The highest permanent human settlement is a mining town in southern Peru called La Rinconada at an altitude of 16,730 feet. Population: 7,000.

he collapsed, unconscious. Fellow climbers hauled him on a makeshift stretcher to base camp—altitude 17,559 feet—where Kamler was waiting with a surgical bed built from rocks. Pasang's face was bloated, his eyes were swollen shut, and he was in a deep coma. Without question, his brain was bleeding and the pressure was building inside his skull. Kamler treated him with oxygen and IV fluid, but there wasn't any more he could do. Brain surgery wasn't an option. Then the other Sherpas began to chant for their friend. Kamler describes it as a rumbling sound that seemed to emanate from the great mountain itself. He began to wonder whether this rhythmic chanting was reaching Pasang, resonating deep inside his head, perhaps harmonizing with his brain waves. If it was really happening, Kamler thought, "this effect might be powerful enough to reverse a shutdown." It may sound outlandish, even preposterous, but he says anything can happen when it comes to the human body and survival. "I have learned not to dismiss this kind of possibility," he writes. "No course in medical school taught me the proper mixture of oxygen, IV fluids, and Tibetan chants to treat a subdural hematoma in below-zero temperatures on a 3-mile-high glacier."

The laws of medicine dictate that Pasang should have died from his injuries. People don't survive similar head bleeds one block from the hospital, let alone on the frozen side of the world's tallest mountain. And yet overnight, Kamler watched Pasang's thready pulse and swollen face slowly return to normal. By morning, the Sherpa opened his eyes. Kamler had no scientific explanation. The oxygen and IV couldn't have been enough. "It was shocking," he says. "We can never fully understand it." He concluded that "the chanting had released an energy within Pasang, a will to live, and this had reversed his decline." It was an unbelievable moment for a doctor. It also led to his conviction about our extraordinary and mysterious ability to survive the impossible. "I realized then," he writes, "that practicing extreme medicine would sometimes mean witnessing and working alongside phenomena I might never understand." Kamler's observation brought me right back to Brian Udell, brutalized by 43 sustained g and clinging to a life raft in the Atlantic.

3. "I Can't Die Tonight"

Pounded by the ocean and on the verge of giving up, Udell says one vivid image appeared in his mind's eye and drove him to keep fighting: his pregnant wife. In October 1993, he had married his true love, Kristi. Now she was a few months pregnant with their first child. As the waves battered his energy and confidence, he imagined a devastating scene: An air force notification officer walking up to the side door of his home in Goldsboro, North Carolina. In this nightmare, Kristi was standing in the doorway, receiving the terrible news that her husband had died in a training accident. It was so unfair, Udell thought. How cruel to Kristi and the unborn baby. *I can't let that happen,* he told himself. So he prayed to the Lord to let him see his wife again and to witness the birth of their child. *I've got too much to do,* he remembers thinking. *I can't die tonight.*

Udell suddenly felt a surge of energy. Summoning all his strength, he made one last attempt to pull himself onto the raft. This time, instead of knocking him off, a gentle wave nudged him to safety. The feeling was incredible. His entire body ached, but now he was out of the cold water and he could take care of himself. No doubt the searchers were already hunting for him. Even if it took hours or days to be rescued, he knew he would be okay. A calm, spiritual peace surrounded him. He looked down at his body, limbs akimbo, and couldn't believe the damage. Everything seemed twisted in the wrong direction. He had broken or dislocated just about every major joint in his body, but in the deadpan style of a fighter pilot, he thought: *Dude, you've got to straighten yourself out.*

Four hours later, a Coast Guard helicopter plucked Udell from the Atlantic. It took a rescue swimmer thirty minutes to disentangle him from the parachute lines and hoist him to safety. When air force investigators arrived at his bedside in the New Hanover Regional Medical Center in Wilmington, North Carolina, one of them stared at Udell and said, "You're not supposed to be here. The human body isn't designed to handle that." It's a marvel of understatement. Udell and his partner Dennis White ejected at almost the same exact moment. The circumstances were almost identical, and yet Dennis was killed instantly. Searchers found his body in the ocean fourteen hours later. Investigators believe he died within a few heartbeats of ejection. The windblast and g-forces ripped him apart, severing his femoral artery. Mercifully, he didn't feel a thing.

When I ask Udell why one survived and the other didn't, he just shakes his head. "I have no clue," he says. "Those things are a mystery." He pauses as the emotions get to him. "Both of us should've been killed." He stops again for a few moments, then points out that he's six foot one and weighed 195 pounds when he ejected. Dennis was taller and skinnier at six foot three and 185 pounds. Maybe that was the difference, Udell speculates. The taller, thinner man was pulled apart by the forces while the shorter, stockier man held together. But it's just a theory.

Like many survivors, Udell doesn't claim any credit for what he did. He's deeply modest about what he endured, almost to the point of seeming shy and uncomfortable. "There's nothing superhuman about me," he says. "I'm a normal guy." During the brief moments between ejection and the automatic deployment of his parachute, he believes his fate was out of his hands. It was up to the Lord and the laws of physics whether he could endure what no one had ever survived before. Once he made it through the g-forces and windblast and was floating down under his parachute, survival was up to him. Military training took over. He knew what to do and how to do it. He had a plan and executed it. But when he couldn't pull himself into the life raft, he believes God gave him one more assist by sending a wave that lifted him to safety. For Udell, survival teetered between the controllable and the uncontrollable, the known and the unknown. On this razor's edge, life and death are determined every minute of every day.

In a sprawling complex north of Phoenix, Arizona, the Goodrich Corporation owns a factory that makes ejection seats. In December 2001, Udell paid a special visit to offer his gratitude. He toured the plant and watched workers carefully measuring gunpowder. The precision of their labor is absolutely critical to a pilot's survival. When an aviator pulls the ejection handles, three hundred different explosive devices must fire in a precise sequence. Too much gunpowder—even just a sprinkle—and the explosion would be too powerful. Your neck would break or your spine would snap. A smidge too little and the ejection wouldn't be fast or forceful enough. You wouldn't clear the tail of the plane, and you'd be cut in half. "I owe my life to you guys," he told the workers. Their attention to detail meant that when he needed to escape from his jet, the rocket seat worked perfectly. Years later, Udell's eyes well up with tears thinking about those men and women in the Goodrich factory. They're part of a long survival chain that kept him alive. The list includes the people who hand-make air force survival rafts in Florida and the Coast Guard helicopter rescuers who winched him from the ocean.

Incredibly, within ten months of the accident, Udell was flying F-15s again and went on to serve two more tours in Iraq. It would have been sooner, he says, but it took a while to get two waivers from the air force. One allowed him to fly with thirteen pieces of metal in his body. The other permitted "laxity" in his knee because he was missing ligaments. At 3:36 PM on September 7, 1995, Udell witnessed the birth of his son Morgan Daniel. In that instant, Udell says, all of his prayers in the Atlantic had been answered. "This is what you fight for. This is what you live for," he says full of emotion. "Pain is temporary. This is eternal."

Humble people like Udell recognize the tension between the manageable and the unmanageable. They share a major responsibility in their survival. Their attitude, experience, and preparation make a critical difference. So do the first responders, nurses, and doctors who forge key links in the chain of life. But survivors also appreciate how much lies beyond their grasp. They recognize that fate and chance often trump our best strategies and tactics. At this crowded intersection of what we command and what we don't, the most effective survivors know when to hold on, when to let go, and as the faithful say, when to let God.

* * *

In my investigation of the secrets of survival, Brian Udell and Ken Kamler opened the doors to the metaphysical. Two highly trained and serious men—a fighter pilot and a physician—showed me that for all the science in the world, sometimes there really are no earthly explanations for who lives and who dies. Sometimes, we must look beyond, and that is where I turn next: to my encounter with a young woman who went out for a bike ride and suddenly found herself locked in the jaws of a lion—a testimony to the awesome power of faith.

Survival Secrets

The Magic Numbers of Staying Alive

In almost any survival emergency, the US Air Force believes two magic numbers can help save your life. In its grueling survival course at Fairchild Air Force Base near Spokane, Washington, the first thing that instructors drill into your head is 98.6. Whatever you do, they say, protect your core body temperature. They don't even bother to attach degrees to the numerals. Their slogan is simple: *Maintain 98.6*. It's the top priority. As we learned in our encounter with Professor Popsicle, cold kills. In a survival situation, everything else—like food and shelter—can wait. Once the air force bangs 98.6 into your brain, there's one other digit it force-feeds. Every student learns the magic number 3. Whether you're an F-15 fighter pilot or a single mom, in a car accident or taking a walk in the park, the number 3 will keep you alive.

The Rule of 3 states that you *cannot* survive:

3 SECONDS WITHOUT SPIRIT AND HOPE
3 MINUTES WITHOUT AIR
3 HOURS WITHOUT SHELTER IN EXTREME CONDITIONS
3 DAYS WITHOUT WATER
3 WEEKS WITHOUT FOOD
3 MONTHS WITHOUT COMPANIONSHIP OR LOVE

Air force training emblazons the number 3 on your mind. They make you memorize the order of the rules so that you will always know your survival priorities and be able to manage your needs. Then they drop you off in the woods, deprive you of food, and run you around till you can't think straight. They teach you which plants to eat and which are poisonous. They show you how to make a slingshot and hunt for dinner. They rough you up and interrogate you like a prisoner of war. And at the end of the seventeen-day course, they hope you never forget those numbers: 98.6 and 3. In the fog of war or the downpour of a storm, those magic numbers keep your priorities clear and help you stay alive.

6

Rescued from the Lion's Jaws

Prayer, Miracles, and the Power of Faith

I'm in big trouble.

Those were Anne Hjelle's first thoughts when she saw a flash of fur over her right shoulder. Mountain biking along a twisting trail in Foothill Ranch, California, she thought she had startled a deer. *If only.* In a streak of speed and force, a creature pounced on her from the brush, knocked her off her bike, and plunged its fangs into the back of her neck. Anne knew immediately it was a mountain lion. She had seen signs posted at the trailhead. The big cats had been spotted in the park, and riders were supposed to be extra careful, but whoever paid much attention to those warnings? The animals are supposed to be afraid of humans. At least that's what Anne had always believed.

"Jesus, help me," Anne cried out. *Jesus, help me.* It wasn't some casual quip. These three words were entirely intentional. "It was a conscious response to my previous thought that I'm in big trouble and I'm going to die," she tells me. As the lion clamped down on her head and tried to drag her off the trail, Anne says she purposefully called out to God.

I join Anne and her husband, James, for a mountain bike ride not far from the scene of the attack. It's a choppy trail, and Anne keeps a close eye on me so I don't end up impaled on a cactus.

She was born in Apple Valley, Minnesota, but she's a California girl at heart. Today, she's wearing black Lycra shorts, a blue tank top, and big sunglasses. Her blond hair is pulled back in a ponytail. Unless you look closely, you don't really notice the scars. She points out that one of her blue eyes droops a little, and she'll need more surgery to get it fixed. But when you see her on her bike—or when she walks into a coffee shop—she's a perfect advertisement for the happy, healthy California lifestyle. On this gorgeous day with a vast blue sky and the Pacific Ocean glimmering beyond the hills, we ride and talk about January 8, 2004, the day a lion tore off her face and the Lord answered her prayers.

The Whiting Ranch Wilderness Park is a fifteen-hundred-acre expanse of canyons, ridges, creeks, and trails in Orange County. It was 4:15 PM, and Anne and her friend Debi Nicholls were cycling on Cactus Ridge Trail, a narrow, twisting track that cuts through scrub and brush. It's a fast, bumpy ride like a bobsled run, and Anne was speeding along at fifteen miles per hour—a good clip when you remember that the dirt trail wasn't much wider than the book in your hands. As she crested a plateau, Anne spotted the furry flash in her peripheral vision, and then something smashed her from the right side. The impact was incredibly powerful. Anne weighed 125 pounds, and she would later learn the lion was 122 pounds. Cougars—as they're also known—take down big prey with a body slam. Anne was thirty years old, five foot four, a personal fitness trainer and a self-described tomboy, but in the animal's clutches she felt like a rag doll. "He's got complete power over you," she says.

On the ground, the lion attacked the back of her neck below her bike helmet and quickly worked its jaws toward her face. She felt its fangs open and close, shifting slightly each time, angling for her throat and the kill. As the mountain lion dug into her neck, her instincts took over, and she struck back, punching the animal in its face but unable to strike its body. Later, in the hospital, doctors found that her hands and knuckles were black and blue from so much fighting.

Without any emotion in her voice, she takes me through the carnage. The lion's two upper fangs broke her nose and punctured her upper lip. The bottom fangs pierced her cheek closer to her ear. "As he clamped down," she says, "I knew that he just basically tore my face off, and I remember thinking to myself, kind of wondering *Do I want to live,* because I knew he just destroyed my face." Once more, she responded instinctively. In her husband's self-defense classes, she had learned that if someone tries to choke you, you're supposed to turn your head sharply and jam your chin down to your shoulder so they can't get a grip on your neck. Anne sensed the lion was going for the death bite—the final attack on her jugular—so she turned her head and body away to protect herself. "That was when he released from my face," she says, "and grabbed me by the front of the throat."

I ask if she remembers what the animal looked like. Could she smell its breath? Did it make any sounds? She laughs. She doesn't recall any growling, and her friend Debi told her the cat was eerily silent throughout the attack. "He's so close range," she says, "there's not a lot that I'm seeing." She'll never forget that the color of its fur was reddish, but when a creature's jaws are literally chewing on your face you really don't have a very good view.

If you bite into an ear of corn, the masticatory muscles in your jaw compress your teeth to exert around sixty-eight pounds of force per square inch. If you clench really hard, you can increase the pressure to 170 pounds. For comparison, a Labrador retriever bites with a force of around 150 pounds. Now consider a mountain lion, which has a bite force of around 940 pounds.* Typically,

* Researchers have found that alligators have the strongest bite in the animal kingdom. The force of a twelve-and-a-half-foot gator named Hercules was measured at 2,125 pounds. The dinosaur *T. rex* is believed to have chomped with 3,300 pounds of pressure.

cougars kill smaller prey with a death bite at the base of the skull, snapping the neck and spinal cord. If that doesn't work, they go for the jugular and windpipe. Their victims usually bleed out or suffocate, and it's over very fast. "Even with the tearing away of my face, what struck me was the strength required for that just to peel back," Anne says. "You think of all these tissues attached to the bone and everything. It wasn't pain. But I remember thinking, *Wow!* It felt like nothing for him to do that."

Anne knew her time was running out, but suddenly she heard her friend Debi screaming incredible obscenities. The profanity—especially the F-word—startled her because she had never heard her religious friend swear. Then she felt Debi pulling hard on her left leg, trying to free her from the beast. In this vicious tug-of-war, Anne was the rope, and she didn't think she would survive. *This is the end,* she told herself. She tried to say good-bye to Debi, who refused to listen. "I just told her, 'I'm never letting go,'" Debi says.

Amid the chaos and struggle, Anne felt completely at peace. Her vision and thoughts grew fuzzy. *Well okay,* she told herself, *I guess that's it.* As she began to black out, she wondered: *How come I'm not seeing my life flash before my eyes or a tunnel of light?* When she awoke some thirty seconds later, the lion was gone and Debi was kneeling near her. A few other bikers had stopped to help, throwing rocks at the animal to scare it away. Anne felt like she was drowning in blood, so she tried to sit up and clear her throat. Then she focused on slow, deep breaths. She wanted to stay calm and maintain control. Even with part of her face hanging down like a loose flap, she found herself in awe of the moment. "That's unbelievable," she said to Debi. "That was unbelievable!" Even today, she can't quite comprehend it. "You know," she says softly, "it's so surreal still."

Mountain lion attacks are incredibly rare. Between four thousand and six thousand big cats roam California, and there have been fifteen maulings and six deaths since officials began keeping records in 1890. The chances of even seeing a lion are pretty slim, but on the day that Anne was ambushed, wildlife officials also discovered the half-eaten body of an avid mountain bike racer named Mark Reynolds, a thirty-five-year-old executive with a sports marketing company. He was attacked on the same trail. The lion later ambushed Anne, who was rushed by helicopter to a hospital in nearby Mission Viejo. Doctors quickly determined that the lion's fangs had nicked her spinal column and sliced within a hair of her jugular vein, carotid artery, and windpipe. They immediately treated forty deep bite marks in her neck and reattached her face using more than two hundred stitches and staples. At the very same time in Whiting Ranch Wilderness Park, trackers found and killed the mountain lion. It was a two-year-old male with reddish fur.

1. "Fear No Evil"

When I started writing this book, I was somewhat skeptical of the role of faith in survival. As I admitted in chapter 3, I've always offered a few prayers during an airplane's takeoff and landing, still I was doubtful of their impact. But as I began to interview survivors around the world, I noticed a remarkable pattern. Overwhelmingly, they shared a belief that God and faith had sustained them through their trials. As many as 75 or 80 percent cited a higher power as an important reason for their survival. Over time, my incredulity proved no match for their conviction. Indeed, I began to feel admiration for their faith and I envied their certitude. Military folks like to joke that there are no atheists in foxholes. When bullets fly, everyone prays. In the trenches of survival, it's the same. Many survivors are true believers. Either they face their crisis with strong faith or they discover it in the crucible. Of course, their beliefs differed in many ways, but they shared a common perspective. God had a plan for them and gave them the strength to overcome. Some, like Anne Hjelle, believe that the Lord actually delivered them to safety. Others maintain that when they were lost or down, God guided them to a better place.

Perhaps the most unexpected spot where I encountered faith was at the Naval Survival Training Institute in Pensacola, Florida. I traveled to the sprawling air station on the Gulf Coast to meet Ray Smith, a guy who knows that the manzanillo or beach apple tree can cause blindness if its poisonous sap gets in your eyes. Smith co-authored the latest edition of *How to Survive on Land and Sea,* an encyclopedia of practical information on what to do if you're stranded in the tropics, desert, mountains, or ocean. He's a genial fellow with a warm North Carolina drawl, dark eyes, and a salt-and-pepper mustache. A paunch protrudes beneath his blue shirt. Once upon a time, Smith says he was an inch taller and twenty-five pounds lighter. As a hard-ass marine drill instructor, he showed young recruits how to kill in hand-to-hand combat. Later in the navy, he spent twenty-seven years on active duty teaching people how to survive. He's seventy now and works as a contractor with the Survival Training Institute. His office is located in a modern glass building on the waterfront of the Pensacola base. Squads of chanting marines jog by in tight formation. I'm hoping Smith will share the tricks of the trade. I begin with a simple question: "What's the secret of survival?" Without hesitating, he answers: "Faith in God."

"Really?" I ask. "Absolutely," he says. "It's a major factor in all survival scenarios." Smith's close friends include some of America's most legendary ex–prisoners of war. Faith is a unifying force in all of their experiences, he says. Indeed, Smith is such a believer that one navy skipper nicknamed him the Chaplain. While editing his survival manual, Smith wanted to start with a verse from Psalm 23: "Yea though I walk through the valley of the shadow of death, I will fear no evil, for thou art with me; thy rod and thy staff, they comfort me." But his publisher said no way. "It's just not appropriate," he was told. Smith insisted and ultimately prevailed. He feels so strongly about faith that he thinks it properly belongs as the first chapter of any book on survival. In military parlance, it's a force multiplier, a factor that significantly increases or multiplies your strength or effectiveness. When you're feeling weak, he says, faith pumps you up. When you're run down, it gives you a boost. When you're discouraged, it lifts you up.

It's a beautiful November day in downtown New York. The Chambers Street subway station buzzes with morning commuters. Stan Praimnath greets me at an underground newsstand. Born and raised in Guyana, Stan is a cheerful and fastidious fellow. He leads me to the mezzanine overlooking the so-called pit, the void where the World Trade Center once stood. On this day, workers are beginning to pour the foundation of the new Freedom Tower. We watch the men moving heavy machinery where almost three thousand people perished. "When I look down in the pit today," Stan tells me, "I see myself lying with a group of dead bones down there." He pauses. "I may never know why the Lord intervened on my behalf. I've stopped many, many days and asked myself, 'Lord, why me?' Of all these men, women, and children? But I stopped and I asked myself one day, 'Stan, why *not* you?'"

On two separate occasions, Stan escaped death in the World Trade Center. After each terror attack, he believes God rescued him. On February 26, 1993, he was on the seventy-ninth floor of the South Tower eating a take-out lunch of Chinese pork chops and rice when a fifteen-hundred-pound car bomb went off in the parking garage below the North Tower. The building shook and the lights flickered, then went out. "I thought I was going to die when the bomb exploded the first time," Stan says.

It was a windy, snowy day, and as the smoke began to rise, his co-workers screamed and ran in every direction, but Stan calmly ate his meal. His supervisor shouted, "Stan, how can you eat at a time like this?" He replied: "Lord, if today's the day I'm coming home, I'm coming with a full stomach." Six people died that day, and 1,042 others were injured.

On September 11, 2001, Stan was standing at his desk on the eighty-first floor of the South Tower. It was 9:03 AM. Some eighteen thousand people had come to work that morning at the Trade Center, and Stan was an assistant vice president in the loan department of Fuji Bank. Seventeen minutes earlier, when American Flight 11 had slammed into the North Tower, he had evacuated to the lobby but returned to his office when he was told to go back to work. He was talking on the phone with a very concerned co-worker in Chicago. "Stan," his colleague was asking, "are you okay?" He remembers looking out toward the Statue of Liberty and seeing the most terrifying sight: A giant gray aircraft with a U on its tail was flying straight at him. Stan felt like he was staring "eyeball-to-eyeball" with United 175. "I cannot verbalize that sound," he says. The Boeing 767 was flying 590 miles an hour with sixty-five people on board, including five hijackers. "I can hear that screeching sound as it's coming toward the Trade Center, coming for me," he says. "I don't think any doctor or any psychologist or anybody would be able to take that sound away from me."

With the United jet just a few hundred yards from the South Tower, Stan dropped the phone and shouted, "*LordIcantdothis-youtakeover.*" Today the eight words still burst from his mouth like one. Lord, I can't do this, you take over. He still blurts out the phrase just the way he did on September 11. There was no time for pauses. He dove beneath his metal desk just as the 767 slammed into the building. Everything between the seventy-seventh and eighty-fifth floors was destroyed. Some six hundred people died immediately or were trapped in the wreckage. "The fact that I was not killed instantly was certainly a miracle from above," Stan says. When he peeked out from beneath his desk, the destruction was unbelievable. The fire sprinklers showered the floor with water, and blue sparks arced in every direction. Impossibly and unmistakably, the wing of the plane was wedged into his door just twenty feet away. Everything else was obliterated, but Stan noticed that his Bible was resting undisturbed on his desk. "The Word of God was not to be destroyed," he says.

Stan cried out again to God. "Lord, send somebody, anybody. I have two small children! I don't want to die. Why am I alone? Send someone, Lord." In that moment, he imagined his wife, Jennifer, and daughters Caitlin and Stephanie. "My greatest fear was, who was gonna take care of all these bills that we had incurred? Who was going to walk our two daughters down the aisle when they get married? That was my greatest concern." Stan hesitates for a moment. His gentle face contorts like he's back in South Tower. "I just want to go home," he says, "just want to see my two girls, just want to hold my wife, know that all is well. That was my main concern. I don't care what happened. I don't care if I lose my leg and lose my life, but Lord, intervene, I want to go home."

Stan crawled through the debris, looking for help or a way out. Everyone was gone—he would discover later that they had all died—and he felt so alone. Then in the distance, he saw a beacon in the smoke and fire. "I see the light," Stan shouted, "I see the light." It was a flashlight in the hands of a man named Brian Clark, a vice president with Euro Brokers on the eighty-fourth floor and a fire warden. Stan believes Brian was his guardian angel, sent by the Lord Himself. "I would never understand why people pray all their lives, not see the results," Stan says, "and here I am calling out on God, this split second, and He sends someone for me." His face beams at the thought. "I'm not a religious fanatic," he says, "and here I am screaming out to God, and . . . in retrospect, there's no call waiting, there's no operator, there's no long-distance carrier." Stan marvels that God answered so quickly. "I call, He heard, and He intervenes. And like I said, I'm not a religious fanatic. I'm a practicing churchgoer, that's who I am."

2. How Religion Helps You Live Longer

For Stan Praimnath, survival isn't a complicated equation. There's no need for different variables like age and extent of injuries. God is the only explanation. Everything else is unnecessary, even irrelevant. If you want to grasp the power of religion in survival, consider this extraordinary fact: *People who go to church regularly live around seven years longer than people who don't.* That's right: seven years.* More precisely, if you go to church once a week, your advantage is 6.6 years. If you worship at church *more* than once a week, your edge increases to 7.6 years, a bonus of one additional year. It's well known that factors like gender and race influence how long people live, but who knew that religious attendance makes a significant difference, too?

Dr. Harold G. Koenig of Duke University Medical Center is one of the pioneers in the field of faith and health. As founder and director of the Center for the Study of Religion/Spirituality and Health, he's written more than thirty-five books and three hundred articles on how people's religious beliefs influence their mental and physical well-being. The fancy name for the field—get ready for a mouthful—is psychoneuroimmunology, or PNI. It's the science of how your mind influences your health and how social and psychological factors like religion affect your

---

*This remarkable statistic comes from a 1999 study conducted by researchers at the University of Texas at Austin.

immune and nervous systems. Dr. Koenig tells me that he's *not* trying to prove the existence of God or anything else supernatural. He simply wants to understand the impact of faith in the lives of 80 percent of the world's population who are involved in organized religion. That's 5.2 billion souls.*

I ask Dr. Koenig about the stunning seven-year statistic. "That has nothing to do with whether God exists or doesn't, whether prayer works or not," he explains. It's based entirely on the fact that religious attendance produces "psychological, social, and behavioral consequences" that help you live longer. "They would work whether God exists or not as long as people behaved like and believed that God existed," he says. It is not clear whether the particular religion matters, Dr. Koenig adds. More research is needed to determine if the effects are the same for Christianity, Buddhism, Islam, Judaism, or any other creed. However, longer life appears to be correlated with the extent to which your faith is integrated into your daily decisions and actions. People with committed religious beliefs tend to have stronger support systems and more solid relationships; they are more likely to follow teachings that reinforce a healthier lifestyle. In short, observant people usually don't smoke, drink, or engage in risky business.

What about people in survival situations? How does prayer help someone like Anne Hjelle or Stan Praimnath? Dr. Koenig replies that belief is the most powerful survival tool in the world. Faith gives you hope that no matter what you're going through, something good can come of it. It also gives you a sense of meaning and purpose that can help you overcome incredible adversity. Almost anything is possible, Dr. Koenig says, when you believe that God loves you, that He has a plan for your life,

---

* Ninety percent of Americans are affiliated with a religion and 71 percent pray at least once a week, according to a definitive study in September 2006 titled *American Piety in the 21st Century.*

that He will never leave you alone and will give you strength to handle your hardships. Faith and religion, he says, empower you with "the kind of strength that nothing else that I've ever seen can give."

So who turns to God in a crisis and who doesn't? Dr. Koenig says people are equipped with an incredible range of psychological strengths. Some are "genetically endowed with this tremendous willpower and ability to overcome regardless of what the barriers are." These people are blessed with a "kind of absolute, single-minded, focused survival instinct." Others simply don't have that genetic and psychological strength. They're more vulnerable emotionally and have fewer inner resources in a crisis. On this continuum of strong to vulnerable, Koenig says, religious beliefs help everyone, but they're especially valuable to those on the weakest end. People who depend most on faith often have "nowhere else to turn" or don't have "the resources available to survive." For these vulnerable people, even when a challenge seems impossible, religious beliefs can sustain them through almost anything. "There's amazing things that you can do," Dr. Koenig says gleefully, with "the creator of the entire universe, of history, and of the future . . . at your side."

I ask Dr. Koenig about his own religious beliefs. "I am a person of faith," the fifty-seven-year-old replies. He believes deeply in the healing power of religion because of his own life experience. As a young man, Koenig was admitted to the School of Medicine at the University of California–San Francisco, one of the most prestigious in the country. But in 1976, his life took a stunning turn after he was expelled because of emotional problems. "I was on the streets of San Francisco for a time," he remembers, "really headed for the gutter." Homeless, he lived underneath buildings, washed with a hose in other people's yards, and used a coffee can as a toilet. "That was quite a fall from grace," he says. Dr. Koenig was raised Catholic but wandered away from faith during this time.

How did he rebound and find his way from homelessness to a professorship at one of the America's great universities? "Part of what got me through is religious faith," he explains. "I tend to be one of those weak people with few resources. I tend to be emotionally sensitive. Things bother me a lot, and I'm probably prone to depression and anxiety." He says that as a "vulnerable creature," he was able to "slowly crawl back to sanity" thanks to prayer. With faith emboldening him, he worked his way up the medical ladder from orderly to nurse to physician to professor. In the midst of this arduous ascent, he developed disabling arthritis with chronic pain syndrome and later prostate cancer. Dr. Koenig says he wouldn't have been able to achieve anything—let alone survive arthritis and cancer—without faith. "It would be hard for me to raise my children and stay married to my wife without religious faith," he adds. "It would certainly be impossible to deal with many of my colleagues without religious faith." The last thought makes him laugh.

Every morning in Elmont, Long Island, Stan Praimnath prays in the shower: "Lord, cover me and all my loved ones under your precious blood and take me to work and bring us back home in peace and safety." It is an entreaty that he makes each day, sometimes more than once. Then he goes to his closet to get dressed and sees the Timberland shoe box on the shelf above his shirts and suits. The carton is wrapped in packing tape, and Stan has scrawled one word in cursive on each side: DELIVERANCE. "That box is watching me," he says. "If ever I get cold for the Lord," Stan has instructed his wife, "I want you to open that box and show me where I was and how the Lord delivered me." The parcel contains a pair of Knapp shoes that cost $110, an undershirt, a flashlight, and some medicine. They are all that's left of his posessions from September 11.

Stan and I are riding a commuter train in a tunnel beneath the Hudson River en route to his office in New Jersey. "All my life I know that if I call on God, He is going to help me," he says. "I'm not a strong person. I'm not big." And yet, he somehow managed to get through the impact zone in the South Tower and make his way down from the eighty-first floor. "There had to be a higher being in play," he says. "Strong as I am, weak as I am, I would never have been able to do this . . . but there had to be some force that I can't explain that was working on my behalf that day."

I ask Stan about the thousands who died on September 11. Surely many of them called out to God, too. No doubt they were also driven by the desire to see their families again. Why did God *not* come to their rescue? Stan ponders the question. "Why me of all these people?" he begins. "All these good men, women, and children. Why me? Job said, 'He does wonderful things that we cannot comprehend.'" Stan stops for a second to let the thought register. "In my heart, there is probably something that is unfulfilled, some mission to be accomplished, and that's why I lived . . . I always say, if ever I see the Lord, I will ask Him, 'Why? Why me, Lord?'"

We walk from the train station into the crisp fall morning. A brisk wind swirls off the Hudson River. We're standing in Jersey City on the plaza of Stan's shining office building. It perches on the water directly across from downtown Manhattan. "When I look across toward where the Trade Center was, all I'm seeing is a building that used to be. Colossal. Giant building. Never thought that it would ever go down." But this is Stan's primary lesson from September 11. It is one that he brings to churches across America. He now works for the Royal Bank of Scotland as an assistant vice president, but he also travels around the country on weekends to preach the word of God. "Strong and mighty as you are, you can go down," he says. "Weak and fallible as we are, we can be strong. That strength we can draw from the Lord." He is beaming now, and his smile seems as bright as the sunshine around him.

Stan's faith may be unshakable, but not everyone is so resolute. If religion prolongs your life, what about the opposite? What if you're struggling with your beliefs? Once more, the answer is stunning. It turns out that if you're wrestling with God, it could kill you. Yes, it's a provocative proposition, but one supported by science. Dr. Kenneth Pargament is a psychologist and professor at Bowling Green University in Ohio. He's spent twenty-five years studying spirituality and how people cope with life's toughest challenges. "For the large majority of people, faith is a potent—if not the most potent—resource that enables them to withstand and in some cases grow from the most critical traumas in their lives," he tells me. For a smaller percentage of people, difficult life events like cancer or accidents can shake their faith and beliefs in profound ways. They wonder: Where is God? Does He still love them? Is He forsaking them? Pargament discovered that struggling with God isn't good for your health. Along with Harold Koenig and other colleagues, Pargament studied 596 people who were hospitalized for a variety of illnesses. They were fifty-five or older, and they reported that they felt unloved, abandoned, or punished by God or believed that the devil's work was responsible for their health problems. Grappling with God put patients "at increased risk of death," Pargament concluded, compared with those whose faith remained strong. More precisely, patients in religious turmoil had a 6 to 10 percent greater risk of dying compared with those who weren't. Pargament even figured out which kinds of struggle are especially unhealthy. For instance, patients who felt alienated from or unloved by God and attributed their illness to the devil were 19 to 28 percent more likely to die during the two-year study period.

Why can tussling with God kill you? Like Dr. Koenig at Duke,

Pargament isn't trying to prove or disprove the existence of a higher power. He just wants to understand the health consequence of belief and disbelief. One possibility, he theorizes, is that people in religious turmoil are socially alienated, which in turn means less emotional and physical support. The consequence is poorer health and even death. Pargament is fascinated by what he calls "the fork in the road." Why do some people grow spiritually in a time of crisis while others struggle and decline? The answer, he believes, lies in the nature and maturity of your faith and how much you've integrated it into your life. Pargament says the key question is: "How big is your God?" By that he means, "Do you have an understanding of God or things sacred that's broad enough to encompass both the good and the bad in life? For some people, they've got a kind of sugarcoated picture of God—God will always be there for them, God will never let anything bad happen to them, that kind of thing. And then when they end up faced with a trauma or tragedy, they don't have anyplace to put it . . . so something has to give." On the other hand, Pargament says, some people believe in a larger God "capable of encompassing both the most positive and the most awful experiences in life. And they seem to be the ones who are more capable of assimilating terrible events, tragedies, and growing and moving on." In tough situations, Pargament says, people with the most mature, integrated, expansive view of God seem to handle their challenges the best while those who worship "false gods"—alcohol, drugs, materialism, or narcissism—tend to fall apart.

Pargament is Jewish, and when he began researching religion in 1975 he says he had an "implicit spirituality." Over the years, his faith has become more "explicit." Today the fifty-eight-year-old sees the world through a wide-angle "sacred lens." That means God and religion aren't the only places he finds a spiritual spark. Nature, music, and loving relationships are also an important part of his religious life. The sacred lens brings everything into sharper focus, he says. "You see life with more color, in some ways more dimension, there's a kind of depth to the vision," he explains. "A tree is not just a tree. It's something that's reflective of deeper value. Time can be something that takes on different meaning. Relationships can take on different meaning. The whole world begins to change when you see life through a sacred lens."

3. Does God Answer Prayers?

In a national park near Akron, Ohio, there's a meandering sandy trail that weaves through massive rock formations known as the Ledges. The moss-covered outcroppings are 320 million years old with steep cliffs and lots of shade for an afternoon stroll. In July 1993, the Reverend Lin Barnett, chaplain of the Akron General Medical Center, was walking through the woods with a female friend and her fourteen-year-old son. At one spot on the path, there's an easy twenty-four-inch step over a crevasse. The boy went first with no problem. Then it was the Reverend Barnett's turn. When his foot touched the other side, he slipped and fell backward, plunging forty-five feet into the narrow gap, landing on solid rock. Incredibly, just a few hours earlier, a fourteen-year-old on a Boy Scout hike had fallen near the same spot and died. Barnett doesn't remember what happened that day—he's got amnesia—but he's been told that he was lucky because on the way down he bounced off the jagged sides of the crevasse, slowing his descent. Two medical residents from his own hospital happened to be hiking near where he landed. They took care of him until emergency crews arrived. Barnett was rushed by ambulance to Akron General, where he spent around ten days unconscious in the ICU. He suffered a closed head injury, the kind where the skull isn't fractured but the brain is traumatized.

Barnett's prognosis was bleak. Doctors didn't think he would survive, and if he ever managed to regain consciousness, he wouldn't be able to walk or talk. Friends and hospital workers prayed for his recovery, and suddenly on a Sunday morning, Barnett says, "I woke up." His physicians were astonished—his recovery defied medical explanation—and the first thing he did was ask for two people to join him at his bedside. He chose them specifically because they believed in the power of prayer. They had both been students in his clinical pastoral education course at the hospital and he had been impressed by their spirituality. "I wanted those two people to come and pray with me and they did," he says. Every night they gathered for scripture and prayer, and sometimes in the morning one of them stopped by for more worship. As the weeks went by, the prayer group followed Barnett from the hospital to a rehabilitation facility and then home to his apartment. At each step, the number of participants grew. "It was very soothing and very humbling that so many people prayed for me," Barnett says. Most evenings, three or four people came to his side to pray for God's healing. "That was such a moving thing for me that they cared that much," he says. "I think that's one reason that I survived in such good shape.

"I'm sure it made a difference in my psyche and my mind," he goes on, "and I don't discount it physically. Because I recovered remarkably well. I was told by one physician here at Akron General that I was a walking miracle. He said, 'You know, you should be dead. If you're not dead, you should not be able to walk and you also shouldn't be able to think and talk.'" Today Barnett still sings in his church choir, but he doesn't perform with the Akron Symphony Chorus anymore. He doesn't think his voice is "pretty" enough. His facial muscles aren't the same, either. He can't whistle—a pastime he loved—but he can still pucker for a kiss. Sometimes he mixes up words or can't find the one he's looking for, and he doesn't enunciate as clearly. "So there's some slight damage," he says, "but my goodness, compared to what happened to me, I just shouldn't be able to do all these things I do."

As Barnett tells his story, I can't help wondering: Did doctors save his life, or did the Lord come to the rescue? Was it medicine or the divine? Or both? I ask him directly: Did God answer all those prayers? "Yes sir," he says, adding, "I can't prove it but that's my faith." With so much prayer around him, he adds: "I can't help but to believe that God intervened." At Akron General, Barnett directs the spiritual care department and helps oversee an innovative program called Prayer Partners. On a simple slip of paper like a prescription pad, patients can request prayers on their behalf. A volunteer contacts five prayer partners, who offer implorations for up to five days. After that period, a patient can make a new request, and the call goes out to five more partners. Between forty and fifty prayer "warriors" participate in the popular program. So does it work? Is there any proof? "Well, that's what we don't know," Barnett answers. "I think it worked for me and I think prayer works. I believe prayer has an effect so we assume it works. We assume it brings people healing, spiritual, physical, and/or mental." But Akron General hasn't done any scientific studies. That's missing the point, he says. "This has to be taken on faith."

It's called intercessory prayer—interceding with God for the benefit of another person. In the world of medicine, it is controversial because it's so difficult—if not impossible—to measure the impact of one person's prayers on another person. Over the years, a number of studies have shown that intercessory prayer can actually lead to positive results. For instance, researchers at Duke University Medical Center found that cardiac patients who received stents *and* intercessory prayer or other nontraditional treatments did better than those with standard stents alone.* Seven different groups around the world from various denominations—Buddhists, Catholics, Moravians, Jews, fundamentalist Christians, and Baptists—offered prayers on behalf of patients in North Carolina. Those who received intercessory prayer or nontraditional treatments had 25 to 30 percent fewer "adverse effects" like death, heart failure, or heart attack compared with those who received the standard treatment alone. But for every study asserting the benefit of intercessory prayer, there seems to be another rebutting it. In April 2006, for instance, researchers divided coronary bypass patients at six hospitals across America into three groups: The first received intercessory prayer for two weeks after being told they would *definitely* get it; the second got prayer for two weeks after being told they *might* get it; the third did *not* receive prayer after being told they might get it. Medical complications occurred in 59 percent of the first group of bypass patients who were certain they were receiving prayer. In the two groups of patients who weren't sure, there were fewer complications: Around 51 percent encountered problems after their bypasses. The study concluded that intercessory prayer "had no effect on complication-free recovery" from surgery. To the contrary, the study found, "the certainty of receiving intercessory prayer was associated with a higher incidence of complications."

So who is right? Does intercessory prayer make a difference? Dr. David Hodge teaches social work at Arizona State University in Phoenix. He noticed that a surprisingly high percentage of social workers pray on behalf of their clients. Hodge wondered if this was a good idea. In social work jargon, is prayer an effective intervention strategy? He examined seventeen studies on intercessory prayer and says the answer seems to be a quali-

---

* A *nontraditional* or *noetic intervention* was defined as a "healing influence performed without the use of a drug, device or surgical procedure." Nontraditional treatments included intercessory prayer, stress relaxation, guided imagery, and healing touch.

fied yes. When you look at all the different data, Hodge says, there appear to be "small positive effects for prayer." How does prayer make an impact? Hodge isn't sure. It could be God or a transcendent force or some other naturalistic mechanism that we don't yet understand. Despite the small benefits, however, Hodge concluded that intercessory prayer should not be considered "an empirically supported intervention" for any problem. In other words, it seems to produce results but it shouldn't replace other proven methods of treatment.

When I ask whether he prays for himself or others, Hodge initially dodges the question. "After conducting this study, I need to be more prayerful," he says. He's too busy to worship regularly, but he adds, "I would like to move in that direction." Later, he follows up to say there is "sufficient research" linking personal prayer and meditation with positive health outcomes and these practices are "currently warranted from a scientific perspective."

Anne Hjelle doesn't need science to prove the power of prayer. She knows it makes a difference. Indeed, she believes other people's prayers literally saved her from her worst nightmares. It was around two thirty in the morning at Mission Hospital in Mission Viejo, California. Anne was coming out of her first surgery to sew up the wounds from the mountain lion attack. In the recovery room, Anne couldn't get the violent images out of her mind. The fangs in her neck. The violence of the tug-of-war. The scene replayed again and again like a horror movie. "The only way I could get that vision to go away is to open my eyes," she tells me. "As soon as I close my eyes again, it replays, replays, I mean it was very intense." Later that morning when she awoke, she was visited by her pastors, who sat by her bedside and prayed. They asked God for "a healing of my mind," Anne says. Sure enough, those terrible nightmares and violent images "went away and never came back." Since that visit from the pastors of the Life Church, she's never had another dream or intrusive thought about the attack. It's the power of prayer, she insists. Her pastors asked God to heal her mind, and the Lord answered.

On the second Sunday after the attack, Anne felt strong enough to go to church. To avoid drawing attention to herself, she slipped in the side door with her husband and her family. The sermon was from Second Timothy 4:17, where Paul writes, "I was rescued from the lion's jaws." Anne remembers a feeling of spiritual awe. Those words had been uttered for thousands of years. *Wow,* she thought, *that is so real to me.* When they sang psalms about Christ's salvation, deliverance wasn't some abstraction anymore. God had literally saved her life. "I am not a crier," she says, "but in church singing some of those psalms, I started crying." Tears streamed down her face. Anne is swift to credit her survival to her friend Debi's remarkable bravery, to the cyclists who threw rocks at the animal, and to the doctors who took care of her. But in the end, she believes, God rescued her from the lion. "I cried out and He came to my aid," she says. "That is really what I believe."

4. In Search of Miracles

The first thing you notice about the little old church in Chimayo, New Mexico, is the parking lot right in front. It's big and wide and there's a blue sign posted on the road that lays down the law:

HANDICAPPED PARKING ONLY

Every inch of curb bordering the asphalt lot is painted a familiar shade of blue. By 10:30 AM on this Sunday morning with bells pealing from the church's twin towers, every spot is full, with another car double-parked behind it. They've all come for the same thing: the holy dirt.

It's hard to believe, but several hundred thousand people travel here every year to visit Chimayo, a village nestled in a little canyon carved by the rushing Santa Cruz River. They arrive in huge tourist buses from as far away as China. Many drive from across America, taking a little turnoff on I-40 and then following a two-lane country road that meanders through the pink sandstone ridges and buttes of the Sangre de Cristo Mountains, whose name means "blood of Christ." During Easter week, pilgrims walk twenty-four miles from Santa Fe, many dragging wooden crosses on their shoulders.

The reason for this endless flow of humanity is simple. For almost two hundred years, people have believed that miracles happen in this old adobe chapel, commonly known as El Sanctuario de Chimayo or the shrine of Chimayo. In guidebooks and brochures, it's called the Lourdes of America after the holy site in southern France where millions of pilgrims travel for healing. That's why the parking lot right in front is reserved for the handicapped only. The church makes it easy for disabled worshippers to get to the altar, and I've come to meet the priest who has sat by the front doors for most of the past fifty years. I want to know about miracles of survival.

Father Casmiro Roca stands only four foot ten. He's a Yoda-like figure: tiny, wise, and irresistible. In the far reaches of northern New Mexico, he's known as the little padre with the bald head, gold wire-rimmed glasses, and olive skin that looks almost crispy from ninety years under the sun. He was born in the village of Mura, Spain, in July 1918. Ten months later, his twin brother died. Casmiro was a sickly child, and life was never easy. During the Spanish Civil War, he says two of his brothers were killed because they were Catholic. Casmiro managed to escape to the mountains. When the fighting was over, his father suffered severe burns in a fire and died after forty days of misery. Then Casmiro came down with what he calls an "extreme" illness. A doctor told him he would only survive if he was "born again" and traveled "someplace you have never been, among people you do not know, and work very hard." At the age of thirty-three, the young priest moved from Spain to America. Not long after, he was asked to start a new parish in Chimayo where the old chapel was collapsing. Father Roca literally moved a mountain to repair the church. Challenging the community to help, he asked his parishioners point-blank: "Do you have faith?" They didn't answer right away, but one day a row of trucks arrived in front of the chapel. When they were done, they had moved 150,000 tons of dirt to buttress the old chapel from washing away into the river. To celebrate the Lord's work, Father Roca served the volunteers a gallon of altar wine in paper cups.

In the sacristy next to the old wooden doors of the chapel, Father Roca chortles as he recounts this story. He says he's never missed saying Mass in Chimayo. He also insists that from the moment he arrived here, he was never sick again. He coughs a few times during our conversation and his breathing seems labored, but he says it's no big deal. "I consider my life as extraordinary," he says in heavily accented English. "Never have I been sick anymore."

A middle-aged man from El Paso, Texas, steps into the sacristy. He greets and treats Father Roca like a rock star. The little priest blesses the cross on the man's necklace and bids him farewell. Then he leads me on a tour of the chapel. We walk to the nineteenth-century gold-leaf altar and then enter a prayer room on the side. In the far back, there's a small chamber with a hole in the middle of the floor. It's called the *pocito,* or little well. On Good Friday of 1810, according to lore, a man named Don Bernardo Abeyta saw some kind of light shining from a hillside that led down to the Santa Cruz River. Digging on the spot where he saw the light, he found a cross that he named the Crucifix of Nuestro Señor de Esquípulas. It was carried back to the church in nearby Santa Cruz, but it vanished overnight and mysteriously reappeared on the hillside in Chimayo. This disappearing act happened two more times, and finally Don Abeyta decided to build a chapel to venerate the crucifix right where he found it.

Since that time, pilgrims have come from far and wide to collect the dirt from the *pocito* where the cross was found. Like holy water from the sacred shrines in Lourdes, Fatima, and Guadalupe, the reddish earth here is believed to possess healing powers. So they dig it up using little plastic hand shovels provided by the church. They carry it off in brown paper bags, plastic ziplocks, and round plastic containers marked BLESSED DIRT sold in the town's souvenir stores. Father Roca ridicules the devotion to the dirt. He laughs that people believe it miraculously replenishes itself in the well. "Shut up!" he says. "I pay for the dirt. Tons of dirt." He shows me the adobe shed across the courtyard where a contractor hauls new dirt every few weeks and locks it up. Several times a day, a church worker collects a bucket to refill the well. "I have no faith in the dirt at all," Father Roca says. "The dirt doesn't cure. It's the Big Boss who cures."

In the long rectangular prayer room that leads to the *pocito,* the walls are covered with images of Jesus, Mary, and the saints. Offertory candles twinkle. One table is covered with scraps of paper. On closer inspection, they are prayers for the sick scribbled on fragments ripped from notebooks, brown bags, and even an ice cream wrapper. *Cure his liver, please,* says one note. *Please help protect the souls of my brother and love,* another says. *Free them from the addiction. Keep them in your hands.* There are also racks of crutches, canes, and walkers that have been left behind by people who felt cured by their visit. I count twenty-six metal crutches on one rack and twenty-nine on another. Father Roca tells me that visitors leave behind so many canes and walkers that he gives them away to charity. "I cannot keep," he says of all the medical equipment abandoned here. "There is no room."

I ask Father Roca about the miracles he has witnessed. His face crinkles with a smile that seems to say: *Where should I begin?* Locking the doors at sundown one day, he saw a car with Texas license plates pull up in the parking lot. A man and woman emerged carrying a girl who was six or seven years old. She was obviously very sick, Father Roca remembers. Her doctors had predicted she would die within a few days. "We need to pray," the man told Father Roca. When the family was finished at the altar, the girl suddenly seemed full of life. The man and woman thanked Father Roca profusely and left in the car. Several days later, they returned. The girl's illness had vanished. Doctors couldn't find any trace of disease. She had gone back to school. A miracle.

Father Roca switches to a new story without even a pause. He describes an old woman who was carried to the altar. She had been unable to walk for years. "Father! Father!" someone cried out from inside the church. "Something happened here!" Suddenly the elderly woman stepped out of the chapel toward the sacristy. She was moving all by herself. She was completely cured. A miracle.

Father Roca doesn't keep records of these extraordinary moments. Nor does he follow up and investigate. In fact, he believes in the "constant miracle" of healing that happens every day at the church. People leave the sanctuary with joy in their hearts and souls. That is another kind of healing, he says. It is a miracle, too.

In 1858, a peasant girl in Lourdes, France, named Marie-Bernadette Soubirous reported eighteen apparitions of the Virgin Mary within less than a year. Since that time, the Medical Committee of Lourdes has declared that sixty-seven official cures or miracles have taken place there. The last was in November 2005 and involved Anna Santaniello of Turin, Italy. She visited Lourdes in August 1952 when she was forty-one years old after suffering from Bouillaud's disease, which left her in a constant state of breathlessness. She had trouble walking and speaking, and was plagued with asthma attacks. Anna also had severe heart disease and acute rheumatoid arthritis. She was carried to the sanctuary on a stretcher and one day later, she managed to walk on her own. After years of physical examinations and reviews, the Medical Committee ruled that Anna had indeed been cured.

Of course, not everyone believes in the miracles of Lourdes. The scientist and skeptic Carl Sagan was fascinated by "spontaneous remissions," the scientific term for inexplicable cases in which disease suddenly disappears. If you combine every type of cancer, Sagan argued, the spontaneous cure rate is estimated between one in ten thousand and one in one hundred thousand. If only 5 percent of the pilgrims who visited Lourdes went to heal their cancer, "there should have been something between 50 and 500 'miraculous' cures of cancer alone," he argued in *The Demon-Haunted World*. In fact, only three of the sixty-seven official cures involved cancer. Therefore, Sagan concluded, "the rate of remission at Lourdes seems to be lower than if the victims had just stayed at home." The chances of a miracle cure in Lourdes are about "one in a million," Sagan calculated. "You are roughly as likely to recover after visiting Lourdes as you are to win the lottery, or to die in the crash of a randomly selected regularly scheduled airplane flight—including the one taking you to Lourdes."*

Caryle Hirshberg of the Institute of Noetic Sciences in

---

* In fact, as we saw on page 58, your risk of death on your next flight is one in sixty million, so by Sagan's logic your chances of a miracle cure in Lourdes are actually sixty times greater.

Petaluma, California, has studied thousands of cases of "remarkable recoveries," a phrase she prefers to *spontaneous remissions.* The word *spontaneous,* she says, implies that these people do nothing to promote their own wellness. "From the patient point of view, that's been a very negative thing," she says. She also believes the word *remission* implies a temporary cure. Hirshberg insists there's nothing instantaneous or temporary about remarkable recoveries. These patients typically go to great lengths to make themselves heal. They search out the best doctors, undergo regular treatment, eat properly, exercise, and take care of their minds and souls. In short, remarkable recoveries aren't as miraculous as they appear when you examine how much effort went into the healing.

* * *

Miracles and the power of prayer may be impossible to verify in a scientific study or measure in a laboratory, but perhaps that's ultimately the point. With all my skepticism, I went looking for empirical evidence that faith makes a difference in survival. I interviewed experts and perused scholarly articles on the power of intercessory prayer. I traveled to northern New Mexico in pursuit of miracles. What I learned definitely influenced my life: I'm more focused on seeing the world through a sacred lens; I'm more engaged in my religious community; and I find myself praying more often. While I still have plenty of doubts, I am absolutely certain of this: Faith is the most universal survival tool, if not the most powerful. At a practical level, religion is a mind-set, a collection of attitudes and behaviors for coping with life. At a higher plane, it is ineffable and mysterious. That is why the faithful don't need science or studies to support their convictions. They don't require empirical proof for the power of God. They just *believe.* And that alone gives them strength to overcome and transcend any adversity.

Survival Secrets

The God Helmet

If you, too, want to feel God's presence, you might want to go looking in a dingy basement in the suburbs of Sudbury, Canada, an old mining community in northern Ontario. There you will find Dr. Michael Persinger's neuroscience laboratory. A professor at Laurentian University, Dr. Persinger always dresses in a three-piece suit with a gold watch chain. His daily routine of research, martial arts, food, and sleep is purposefully repetitive and dull, he explains. The steady sameness of his ritual leaves his mind fresh and ready for new ideas.

In his underground hideaway, complete with rat, bear, and human brains stored in Tupperware, Persinger's proudest possession resides in Room C002, a soundproof chamber that he once called Mohammed's Cave after the shelter on Mount Hira where Muslims believe their prophet received his first divine revelations. Persinger's prized object is a yellow helmet rigged with magnets and wires that was created by his longtime colleague Dr. Stanley Koren. In a previous incarnation, it was used for riding snowmobiles in this frigid part of the world. Now it's known as the God Helmet.

For twenty-five years, Persinger has blindfolded people, strapped on the yellow headgear, and exposed them to weak electromagnetic fields similar to those generated by cordless phones and computer monitors. He's tested believers, atheists, mystics, even a clairvoyant, and 90 percent describe experiencing what he calls a "sensed presence." If you're a Christian, that means you may feel the closeness of God

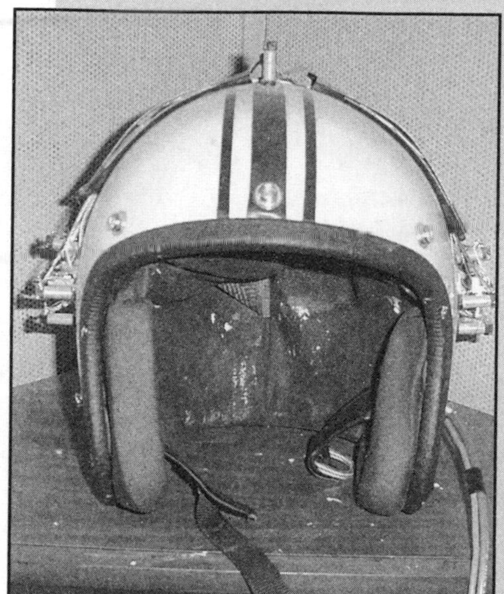
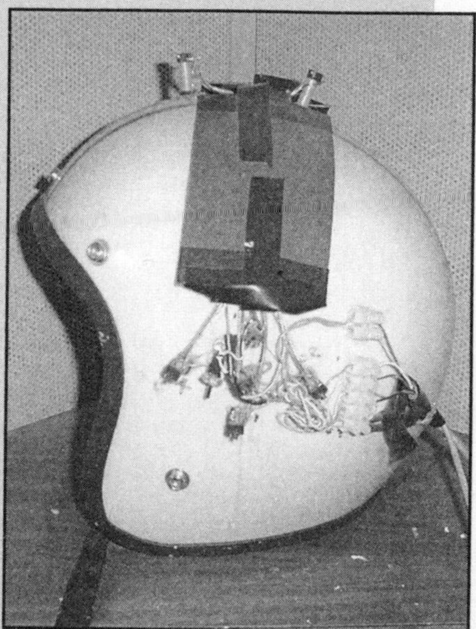

Compliments of the Neuroscience Research Group, Laurentian University

The God Helmet.

or even see an apparition of Jesus or Mary. You may also bask in the warm glow of hope, a sense of significance beyond yourself, or even some kind of force that is guiding or reaching out to you. A handful of people detect ghosts and demons, and one person insisted the devil himself had invaded the chamber. If you're a nonbeliever, you may feel a kind of detachment, as if floating. No matter how you describe it, Persinger says, it's all the same: "The human brain is the source of all experience."

I ask Dr. Persinger if this controversial research with the yellow helmet can help explain why so many people in survival situations feel the presence of God. He cuts me off with a burst of answers. In car crashes, earthquakes, and other survival situations, people experience "maximum hypervigilance." That means the right hemispheres of their brains are remarkably active. As a consequence, they're likely to experience religious thoughts or feel a presence. Pain stimulates these reactions. So does a lack of oxygen. The more activity in your right hemisphere—the part of your brain that fires under stress and distress—the more likely you'll be aware of some kind of meaningful entity in close proximity. These sensations of a higher power, he says, can inspire people to perform incredible physical acts of strength. And, he says, these thoughts of God or angels can lead people to amazing feats of courage and sacrifice.

Working with his loops of wire and magnetic fields, Dr. Persinger has mapped out the specific regions in the right hemisphere of the brain where he says God lives. No matter your religious affiliation or level of belief—or even disbelief—your brain reacts almost identically. What varies is how you describe the sensation. Some people use the language of Christianity or Islam. Others use the vocabulary of Buddhism. Others use the vernacular of the New Age. Your experience, Dr. Persinger says, depends on the activity in the right frontal areas of your brain. During the day, the right hemisphere is less active, but at night—while you're sleeping—it takes over. This is where both creativity and dreams originate. It's also the place for muses, spirits, and all things otherworldly. This is God's headquarters, Dr. Persinger says, and if you're fighting for your life in a survival situation, the right side really lights up. The point of his research, he insists, isn't to prove or disprove the existence of a higher power. The God Helmet simply helps pinpoint the spark in your brain where those spiritual, mystical thoughts are born.*

---

* In April 2005, scientists in Sweden reported that they were unable to duplicate Dr. Persinger's findings. Indeed, they wondered whether he had discovered any real effects or if he had subtly influenced his subjects into believing they were having spiritual experiences. Persinger rejects the Swedish findings, insisting the people in his studies believed they were participating in relaxation experiments and didn't know the true purpose of his research.

7

The Dancer and the Angel of Death

HOW DID ANYONE SURVIVE THE HOLOCAUST?

Edie Eger danced on the day they murdered her parents.

She had no choice, really. Small and skinny, Edie was only sixteen years old when she was deported to Auschwitz-Birkenau, the Nazi extermination camp in southern Poland. Her father and mother were taken away immediately to the gas chambers. On Edie's first night in the camp, Dr. Josef Mengele, the infamous physician, sent his guards out looking for entertainment. Edie was a gifted ballerina, and some of the other prisoners pushed her forward to dance. As a little girl, she had performed for the prime minister of Hungary when he visited Kassa, her town on the banks of the Hernad River. Now she was alone in a room with Dr. Mengele and one of his aides. Her life depended on her artistry. Outside, a band of prisoners played "The Blue Danube," the waltz by Johann Strauss. Inside, Edie closed her eyes and pretended that she was dancing to Tchaikovsky's *Romeo and Juliet* in the famed opera house in Budapest. As her body began to move, her imagination transported her from the barbed wire, smokestacks, and guard dogs. She was gliding across that great stage and twirling in her lover's arms. For a moment, she escaped to a place of perfect beauty. Today her green eyes well up with tears. "I wanted to live so much," she remembers. When her dance ended, Dr. Mengele reached out with a scrap of bread. It was a reward. Little Edie had survived her first day at Auschwitz.

Sixty-three years later, Dr. Edith Eva Eger lives as far as one can imagine from that terrible time. Her home is nestled in an affluent neighborhood near the top of Mount Soledad in La Jolla, California, and her windows look down on a shimmering curve of Pacific coastline. With a big smile and a hug, she throws open her door to a visitor. She still goes by the name Edie, and her Hungarian accent is the only evidence that she immigrated to the United States fifty-nine years ago. She's five foot four, but her charisma is outsize. Her hair is blond and stylishly coifed, and she wears a black turtleneck beneath a brightly colored Escada tunic. Time has taken two husbands, but she has managed to carry on, raising three children and doting on five grandchildren and one great-grandchild. She leads me into her immaculate living room with pristine white furniture and carpet. A graceful bronze statue of a twirling ballerina stands on a pedestal in the corner. Troops of tiny porcelain dancers posture in glass cases against walls that are decorated with colorful paintings by Hungarian artists. This is Edie's sanctuary, where she works as a psychologist helping patients with what she calls "emotional survival." She has spent this morning counseling a twenty-five-year-old woman with chronic pain who wants to commit suicide.

When I ask about her earliest memories in Hungary, Edie says she was always lonely and sad. She doesn't remember laughter or joy. Cross-eyed and "an ugly duckling," she felt like an outcast. Her sister Klara was a violin prodigy, a "superstar" who played Mendelssohn at age five. Her other sister Magda was beautiful, blond, curvier, and vivacious. Edie spent her girlhood "feeling like an orphan, the odd one out." She focused on dancing and gymnastics, hoping some day to compete in the Olympic Games. Without her even realizing it, she says, those lonely times were critical in "developing my inner resources." In a way, unknowingly, she was readying herself for "what was to come."

Between May and July 1944, Hungarian authorities deported 437,402 Jews. Crammed onto boxcars, almost all of them were delivered to Auschwitz-Birkenau. Within moments of stepping off the train, Edie found herself standing in front of Dr. Mengele, a "very scary" and intense man with cold eyes. As the boxcars disgorged onto the platform, Dr. Mengele selected who would go straight to the gas chambers and who would work as slaves in the camps. On that day, he flicked his hand and Edie's mother was dispatched to her death. This was Edie's introduction to what she called the Finger Game. With a simple wave, the Angel of Death sentenced nine out of ten of arrivals to the poison of Zyklon B in the gas chambers. With another gesture, he sent some to the quarantine block for registration. In this first selection, Edie and her sister Magda were spared. That night, she asked how soon they would see her mother. A Polish prisoner pointed to the fiery smokestacks in the distance and said, "She's burning. You better talk about her in the past tense." Later when Edie stood in line for a prisoner identification number, a guard took one look at her and announced it wasn't worth wasting ink to tattoo her left arm because she would die in the gas chambers so soon.

The guard was wrong. Edie and her sister would endure seven months in Auschwitz witnessing brutal beatings and cannibalism in the barracks and five more months as slave laborers in Austria. How did she survive? "I'm *not* a strong woman," she tells me. "I'm a woman of strength." With no chance of rescue from the outside, she realized she would have to find help within herself. She made it through, she believes, because "the lonely child got her inner resources developed." She was tough enough—resilient enough—adaptable enough—to keep going no matter what happened. She knew there was no way to defeat the SS guards. Nor was there any hope of escape. So she came up with a third way. "When you can't fight or flee," she says, "you flow." Survival, in other words, meant adapting to each new day and its unimaginable horrors. This kind of flexibility certainly didn't make her invulnerable to a firing squad or impervious to typhus. But to the extent that she had any control at all over her fate, Edie believes her adaptability and flow gave her a sliver of a chance.

Over the years, some have called Edie "the Anne Frank who lived." She's especially proud of this description because it honors the memory of a rare and sensitive girl. Like Anne who hid for two years in the secret annex of her father's office in Amsterdam, Edie says she spent each day focusing on making choices, no matter how small, about her future. Aching from starvation, she remembers spending the longest time choosing which blades of grass would be the best to eat. Some were thicker with more "meat" than others. A few handfuls staved off hunger for an hour or two. Each decision—no matter how insignificant—felt like it reaffirmed the possibility of the future. *If I can survive today,* she told herself, *tomorrow I'll be free.* Her greatest strength, she believes, was her unshakable conviction that "everything is temporary and I can survive."

On May 4, 1945, Edie weighed around forty pounds and was so sick that the camp guards thought she was dead. Her body was thrown onto a pile of corpses that were supposed to be buried in a forest. Liberating American soldiers arrived just in time, and one of them noticed a hand moving amid all the bodies. Edie was rescued and brought back to life in an Austrian hospital. Her parents were dead but her sister Magda had also survived the camps, and her older sister Klara was still alive after hiding in Budapest. For 20 years, Edie never spoke of the Holocaust or told anyone about Auschwitz. She says she wanted to run away from the past and make a new life in America.

In a quiet restaurant down the hill from her home, Edie says that her survival strategies in the death camps translate to everyday life as well. Whether you're in chronic pain or a survivor of domestic abuse, she believes, you can still make choices to live each day, squeeze the most from limited options, and imagine the future. As they were led by guards from the train at Auschwitz, Edie's mother told her: "No one can take from you what you put in your mind." It may sound self-evident, but Edie believes so many of her patients have lost sight of their power to make choices. What we really need to survive—breathing, sleeping, eating, eliminating—is very different from what we *think* we need, she says. People think they need approval. They think they need to be right. They think they need to be perfect. But none of that is essential to survival. It's just dessert. For lunch, Edie has ordered salmon, broccoli, and chocolate cake. With each course, she fends off a waiter who tries to clear her dishes. Edie is very purposeful about every morsel. Six decades after Auschwitz, she still can't leave a single scrap on her plate.

1. How Did Anyone Survive an Industry of Death?

Nine million Jews lived in Europe in 1939 as World War II erupted. By 1945, six million had been slaughtered. In the most basic arithmetic, three million European Jews managed to escape the Nazi Final Solution. They fled. They hid. They resisted. The survival rate varied greatly from nation to nation, but overall only 33 percent of the Jews living in Nazi-occupied or -dominated Europe survived Hitler's campaign of mass extermination. The survival rate for children was far lower: seven percent made it through the Holocaust.

The Nazis built "an industry of death," as Charles Krauthammer has written, that spanned a continent "with railways, death camps, gas chambers, and crematoria. An industry whose raw material was Jews and whose product was corpses." Amidst this massive and systematic attack on Jewish existence, how did anyone survive? I venture into this territory with humility and awe. First, there is the enormity of the genocide and the fundamental problem of grasping, let alone describing, the evil that befell so many millions. "Survivors feel exasperated and helpless when others who have not the slightest idea what their experiences were like hold forth about what these experiences were all about, and what their real meaning is," the psychologist and Holocaust survivor Bruno Bettelheim has written.* In that sense, it may be both unwise and impossible to generalize about who survived and why. Second, there is the question of accuracy and truth. "Those who have not lived through the experience will never know," Elie Wiesel has written. "Those who have will never tell; not really, not completely. The past belongs to the dead, and the survivor does not recognize himself in the images and ideas which presumably depict him." No single chapter can possibly fathom the reasons why some lived when so many perished. "How does one handle this subject?" the historian Terrence Des Pres has asked. "One doesn't; not well, not finally. No degree of scope or care can equal the enormity of such events or suffice for the sorrow they encompass."

Of course, *Holocaust* and *survivor* are two words that are inextricably intertwined. If one seeks to understand who survives, one must confront the Holocaust. It is the darkest chapter in history, but it raises fundamental questions about the human capacity to endure and overcome. Did the survivors share common traits and strengths? Did those who lived behave differently than those who perished? After liberation, how were they able to build their lives and go on?

As you can imagine, there are many theories about what

* Born in Austria, Bettelheim was imprisoned in 1938 in the Dachau and Buchenwald camps, where he worked as a doctor responsible for the prisoners' mental health. His release in 1939 was purchased, an option that was available prior to the outbreak of the war.

might have made a difference between life and death. In November 2002, for instance, a psychologist at the University of British Columbia reported that the color and shading of a person's hair, eyes, and skin made a meaningful difference in his or her likelihood of surviving. Examining what he called "lethal stereotypes" about Jewish appearance, Peter Suedfeld set out to explore whether people with dark hair and eyes were captured and killed because they "looked Jewish" while others survived because they "did not look Jewish." After questioning 131 survivors around the world, Suedfeld demonstrated that "having light-colored hair, eyes, or both was a positive survival characteristic for Jews during the Holocaust." In other words, Jews with more Aryan features had a better chance. Specifically, 39 percent of his survivor group had light eyes (blue, gray, hazel, or green) versus only 27 percent in a comparison group of Jews in North America who weren't involved in the Holocaust. In addition, 34 percent of the Holocaust survivors had light hair (blond, red, or light brown) versus 25 percent in the comparison group. Why were hair and eye color so important in survival? Suedfeld theorized that lighter-featured Jews might have been able to pass as Gentiles and hide more easily. These Aryan physical attributes may have also made them more "sympathetic" in the eyes of the Gentile population, Suedfeld speculated, and may have even encouraged "helping behavior." In addition, "the persecutors themselves, exalting the Aryan ideal of light hair and light eyes, might have been less rigorous in their dealings with people who approximated that ideal." In other words, perhaps the Nazis were less inclined to murder people who looked familiar.

It certainly makes sense that physical characteristics made a difference, but what about psychological traits? Here again, the range of opinion is vast and contradictory. Many experts and survivors have contended that toughness, ruthlessness, and selfishness were essential qualities for survival. "The worst survived," wrote Primo Levi, who endured eleven months in Auschwitz. By that he meant "the selfish, the violent, the insensitive, the collaborators . . . the spies . . . that is, the fittest." Levi concluded: "All the best died." Viktor Frankl, the legendary psychiatrist and Holocaust survivor, added: "On average, only those prisoners could keep alive who . . . had lost all scruples in their fight for existence; they were prepared to use every means, honest and otherwise, even brutal force, theft, and betrayal of their friends, in order to save themselves. We who have come back, by the aid of many lucky chances or miracles—whatever one may choose to call them—we know: the best of us did not return."*

Could this be true? Did the finest perish and the worst escape? There is certainly no consensus on this point. Indeed, many argue the exact opposite. They contend that altruism, compassion, and selflessness made the difference between living and dying. In this view, survivors formed critical alliances of compassion that improved their chances. Helping one another—sharing rations, tending to the sick, communicating vital information—was the only way to live another day. Still others have tried to analyze the phases of survival in the camps. For instance, Terrence Des Pres has argued that in order to survive, prisoners needed to be able to overcome the shock or "initial period of collapse" when they first encountered the horror. Des Pres described this as a choice prisoners made each time they woke up in the morning. Every day required "psychic acts of *turning*, from passivity to action, from horror to the daily business of staying alive—as if one turned one's actual gaze from left to right, from darkness to possible light." Des Pres insisted, "There was no other way, and to become a survivor, every inmate had to make this turn." Of course, this kind of adapt-

* From 1942 to 1945, Frankl was imprisoned in different camps including Theresienstadt, Auschwitz, and Türkheim.

ability was impossible for many prisoners. "For them there was neither luck nor time," Des Pres wrote. These were the "walking dead" who starved, fell sick, "or stumbled into situations which got them killed."

Bruno Bettelheim argued that the will to live—"a powerful determination"—was absolutely critical. "Once one lost it," he wrote, and "gave in to the omnipresent despair and let it dominate the wish to live—one was doomed." And yet, Bettelheim believed it was "dangerously misleading" to focus too much attention on the individual's role in survival. Interned in Dachau and Buchenwald from 1938 to 1939, Bettelheim attacked the "comfortable belief that the prisoners managed to survive on their own." He argued that "the harsh and unpleasant fact of the concentration camp is that survival has little to do with what the prisoner does or does not do." Until the Nazis were seriously weakened by Allied bombing and defeated at Stalingrad, "not more than a dozen or so of the many millions of concentration camp prisoners managed to survive by their own efforts—that is, to escape from the camps and get away with it before the Allied forces triumphed. All others, including me, survived because the Gestapo chose to set them free, and for no other reason." He concluded that "the few instances of active fighting back—incredibly few, given the millions of prisoners involved—are therefore immaterial to the question of survivorship."

Toward the end of my research on the Holocaust, I decided to make a pilgrimage to New York City to meet a man I hoped would be able to reconcile the conflicting theories about who survived. When Elie Wiesel was only fifteen and a half, a guard in Auschwitz tattooed his left arm with the number A-7713. From that point forward, his identity was erased. His captors never uttered his name again. "I thought when I entered [the camp] I would not survive," Wiesel tells me in his office overlooking Madison Avenue. He is eighty years old now and a nimbus of gray hair encircles his head. His eyes are dark and his brows are bushy. There is an unmistakable aura around him. He is wise, even holy, and yet totally pragmatic.

In his own writing, Wiesel has hinted at different reasons for survival. On the one hand, he has argued that solidarity with fellow prisoners made a great difference. Upon arrival in Block 17 at Auschwitz, he remembers a young Pole offering critical advice: "Let there be camaraderie among you. We are all brothers and share the same fate. The same smoke hovers over all our heads. Help each other. That is the only way to survive." But there were other tactics and strategies, too. "We never ate enough to satisfy our hunger," he writes in the new translation of his extraordinary memoir *Night*. "Our principle was to economize, to save for tomorrow. Tomorrow could be worse yet." Above all, Wiesel says he survived because of his father Shlomo, a shopkeeper from Sighet, Romania. "As long as my father was alive, I wanted to live," he tells me. "I wanted to keep my father alive." Wiesel and his father took care of each other during their imprisonment in various camps. By January 1945, Shlomo "was worn out" from dysentery and fever. Unable to move from his cot, he shivered and gasped for breath with saliva and blood "trickling from his lips." As Wiesel tried to comfort his father, the barrack leader told him: "In this place, it is every man for himself, and you cannot think of others. Not even your father . . . Each of us lives and dies alone. Let me give you good advice: stop giving your ration of bread and soup to your old father. You cannot help him anymore. And you are hurting yourself. In fact, you should be getting *his* rations . . ." This terrible admonition was the diametric opposite of the young Pole's advice encouraging camaraderie.

That night, Wiesel took care of his father, "leaning over him, looking at him, etching his bloody, broken face into my mind." Then Wiesel climbed into his bunk to sleep. When he awoke, his father was gone and another sick person was lying on the cot. "They must have taken him away before daybreak" to the crematorium, Wiesel thought. "I did not weep, and it pained me that I could not weep. But I was out of tears." From that moment forward, he lost all motivation to go on. "Suddenly, the evidence overwhelmed me: there was no longer any reason to live, any reason to fight."

After his father's death, he says, "I did nothing. I simply waited to die." Indeed, he can't explain what kept him alive until the liberation of Buchenwald in April 1945. "It was really luck," he says. In his memoir, he elaborated: "I don't know *how* I survived. I was weak, rather shy; I did nothing to save myself. A miracle? Certainly not. If heaven could or would perform a miracle for me, why not for others more deserving than myself? It was nothing more than chance."

In his private office surrounded by photos, diplomas, and awards from around the world, I ask Wiesel if he believes one Holocaust survival theory over another. His reply surprises me. "All are right and all are wrong," the Nobel laureate says. It would take millions of theories to explain who lived and died, and Wiesel doesn't believe in generalizations. No single argument can account for so many people in such dire and different circumstances. The Holocaust is the exception to every rule, he says. Then he leans forward to share a story about his friend the Dalai Lama, the exiled spiritual leader of Tibet. Many years ago, the Dalai Lama asked Wiesel how the Jews had managed to survive two thousand years after their expulsion from the Promised Land. Wiesel said that the answer could be found in one book: the Bible. The Jews survived, he explained, because they shared a passion for learning and a rare solidarity as a people. These twin pillars sustained the Jews through the centuries until the Holocaust. Then, "all the theories explode." In the end, he believes there is really only one true explanation for who survived. "Chance," he whispers. "It was chance. That is all."

When our meeting ends, Wiesel stands up to bid farewell. He straightens his tie and blazer. He shakes my hand and apologizes for hurrying away. The man once persecuted by Nazis and known only as prisoner number A-7713 has a pressing appointment. He is off to meet with Chancellor Angela Merkel of Germany to talk about the future of Israel.

The Hebrew word for "luck"—*mazal*—is an acronym composed of three different words. *Makom* means "the place"; *zman* means "the time"; and *la'asot* means "the deed." Combined, *mazal* consists of the right place, the right time, and the right action. We'll explore more about luck and chance in the next chapter, but when you ask Holocaust survivors how they made it out alive, most answer that it was *mazal,* plain and simple. You can put forward all the theories in the world, but in the end, they were lucky. They were in the right place at the right time and they did the right thing. Sure, some of them were able to go on because of personal qualities like tenacity and resolve. But given the magnitude of the murder and mayhem, most believe their survival was a matter of chance. In one survey, 74 percent of Holocaust survivors said luck was the main factor in staying alive, while 27 percent credited their coping skills.

Two leading historians of the Holocaust, Henry Friedlander and Sybil Milton, isolated many different ways that luck manifested itself. For instance, a prisoner's minuscule chance of survival depended to some extent on the concentration camp itself. Some were actually deadlier than others. At Belzec and Chelmno, "only five persons are known to have survived; from Sobibor and Treblinka, where revolts took place, less than one hundred survived the war." The "pure chance of geography" also made a huge difference in survival. Very few Jews in Bulgaria, Denmark, Italy, or Romania were killed, while almost every Jew in Holland and Poland perished. The luck of timing was an influential factor, too. "Those deported later in the war had a better chance to live," Friedlander and Milton write, "since they had fewer years to spend in the camps and conditions improved slightly as the Germans needed labor for their war industries." In Auschwitz, for example, very few Slovakian Jews imprisoned in 1942 lived, while a greater number of Hungarian Jews who arrived in 1944 survived. The luck of demography was also significant. "Entire categories of people were automatically sent to the gas chambers: Children, mothers with small children, old people and invalids," Friedlander and Milton explain. Jews who were fit for hard labor—healthy, tall, and strong—had a better chance of surviving. Those with special skills—like calligraphy—were more valuable and kept alive by the Nazis. German-language ability helped, too. And "Jews from Poland or Slovakia, used to the cold climate, had a far better chance to survive than Jews from Salonica, who were used to the mild climate of the Mediterranean."

Chance alone determined many of these factors. There was no time, Friedlander and Milton contend, for Jews to get stronger, learn skills and languages, or adjust to harsh weather. They conclude: "One technique for survival, however, could be learned: the will to survive. Those determined to survive, and willing to make all the needed adjustments through compromise and adaptation, had a chance to survive if all other factors—luck and skills—were also in their favor."

2. The Surprising "Success" of the Survivors

Fifty years after the Holocaust, the Israeli government conducted an extensive survey to determine how many survivors were still alive. The study in 1997 concluded that between 834,000 and 960,000 survivors were living around the world. The greatest number—between 360,000 and 380,000—resided in Israel. Between 140,000 and 160,000 lived in the United States; 184,000 to 220,000 were spread across the former Soviet Union; and 130,000 to 180,000 were dispersed throughout Europe. How did these men and women handle life after genocide? According to conventional wisdom, many suffered from so-called Concentration Camp Survivor Syndrome. They were terribly traumatized and afflicted with serious psychological problems like depression and anxiety.

In 1992, a New York sociologist named William Helmreich turned this conventional wisdom upside down. A professor at the City University of New York, Helmreich traveled across America by plane and automobile to study 170 survivors. He expected to meet men and women who were chronically depressed, anxious, and fearful. To his surprise, he found that most survivors had adapted to their new lives far more successfully than anyone thought. For instance, despite a lack of higher education, the survivors did very well financially. About 34 percent reported earning more than fifty thousand dollars annually. The key factors, Helmreich concluded, were "hard work and determination, skill and intelligence, luck, and a willingness to take risks." He also found their marriages were more successful and stable. About 83 percent of the survivors were married compared with 61 percent of American Jews of similar age. Only 11 percent of the survivors were divorced compared with 18 percent of American Jews. In terms of mental health and emotional well-being, Helmreich found that survivors made fewer visits to psychotherapists than did American Jews.

"For people who suffered through the camps, simply being able to get up and go to work in the morning would already have been a significant accomplishment," he wrote in his book *Against All Odds*. "That they did well in their chosen professions and occupations is even more remarkable. The values of perseverance and ambitiousness and optimism that typified so many survivors were clearly ingrained in them before the war began. What is interesting is how much they remained part of their worldview after it ended." Helmreich theorized that some of the traits that helped them survive the Holocaust—like flexibility, courage, and intelligence—may have contributed to their later success. "That they lived to tell the tale was, for most, a matter of chance," he writes. "That they succeeded in rebuilding their lives on American soil was not."

Helmreich's thesis was controversial, and he was attacked for diminishing or discounting the deep psychological damage of the Holocaust. But he rebuts those critiques, noting that the "survivors are permanently scarred by their experiences and deeply so. Nightmares and constant anxiety are the norm in their lives. And that is precisely why their ability to simply lead normal lives—getting up in the morning, working, raising families, taking vacations, and so forth—makes the description of them as 'successful,' fully justified."

In his one-on-one interviews and a large-scale random survey of Holocaust survivors, Helmreich identified ten characteristics that accounted for their success in life: flexibility, assertiveness, tenacity, optimism, intelligence, distancing ability, group consciousness, the ability to assimilate the knowledge of their survival, the capacity to find meaning in life, and courage. *All* of the Holocaust survivors shared *some* of these qualities, Helmreich tells me. Only *some* of the survivors possessed *all* of them.

So I ask: Which of these ten traits is the most important? And by extension, which tools are most critical to survival in everyday situations, not just the extremes of World War II? "The gift of intelligence," he replies. "Thinking quickly. Brains accompanied by common sense." This kind of basic intelligence—different from book smarts or IQ—enables people quickly to size up situations, break down and analyze problems, and make good decisions.

Helmreich stresses that the Holocaust survivors were just like the rest of us. They weren't exceptional in any particular way. "The survivors were not supermen," he writes. "They were ordinary individuals before the war, chosen by sheer accident of history to bear witness to one of its most awful periods . . . The story of the survivors is one of courage and strength, of people who are living proof of the indomitable will of human beings to survive and of their tremendous capacity for hope. It is not a story of remarkable people. It is a story of just how remarkable people can be."

3. The Silent Secrets of the Survivors

No one made much of a fuss over Holocaust survivors when Rachel Yehuda was growing up in the late 1960s and early '70s in the suburbs of Cleveland, Ohio. Many of her friends had Jewish parents who emigrated from Europe, where they had witnessed—and escaped—all kinds of horrors. Except for their accents, they seemed perfectly normal, like anyone's mom and dad. Yehuda went off to college in New York City and then graduate school in Amherst, Massachusetts, where she earned a doctorate in psychology and neuroscience. As a postgraduate at Yale, she began to work with Vietnam veterans suffering from posttraumatic stress disorder (PTSD). The more she learned about the psychological ravages of war—anxiety, depression, nightmares, and flashbacks—the more she found herself comparing the vets to the Holocaust survivors who looked so different. In her memories of home, the survivors seemed to lead normal lives. Engaged with family and friends and active in the community, they appeared to have moved on from the terror of World War II. Still, Yehuda wondered: Maybe they also suffered from the psychological effects of so much death, destruction, terror, and loss, but perhaps they had nowhere to go for treatment or had their own reasons for keeping silent.

Yehuda decided to study the Holocaust survivors and their "biology of survival." Were they somehow more resilient than the Vietnam vets? Or did they have deep scars, too? She knew one place to go looking—where the truth couldn't hide—was in their urine. By measuring the amount of stress hormones they produced, Yehuda thought she would be able to compare survivors with vets. Her goal was to recruit a scientifically representative sample of survivors from across the country. So in 1990, she asked her parents if she could set up a makeshift laboratory in their basement. Then she approached the local historical society and the survivors' network for help. Soon she was marching around Cleveland Heights and University Heights, trying to collect urine from survivors who were sometimes wary of her request. A vibrant and persuasive woman, Yehuda smiles when she remembers those conversations on doorsteps across Cleveland.

"Why do you want to know about what happened fifty years ago?" an elderly man asked in a thick accent. Yehuda calmly shared her interest in the Holocaust and PTSD. She explained that it was important to learn how survivors adapted to trauma and whether they ever experienced—or still suffered from—psychological problems.

"I don't believe you can learn what is in my heart from my urine," a survivor responded.

Yehuda was fascinated by the reluctance to part with their waste and believes they were genuinely concerned about what it might reveal. "Maybe you should let me try," Yehuda prodded gently. "I'm only asking you for something you were planning to throw out anyway." Once they understood her mission, most of the survivors chose to help. So Yehuda carried their precious urine samples back to her parents' home, where she had cleared out a spare freezer.

For years the conventional wisdom about PTSD had posited that psychologically disturbed Vietnam vets would have *higher* levels of stress hormones like adrenaline and cortisol. The most troubled vets were supposed to be the ones pumping large quantities of these chemicals. In the simplest terms, they were traumatized and flooded with stress. But Yehuda had discovered something quite different. Although adrenaline levels were indeed higher as predicted, cortisol levels were actually lower in Vietnam veterans with PTSD. It was a stunning finding. Yehuda speculated that cortisol acted in some way to help the body recover from the physical effects of stress, like a lubricant for the mind and body. It followed that if a veteran's cortisol levels were *lower* in very stressful situations, his body wouldn't recuperate, putting him at higher risk of disorders like anxiety, insomnia, stomach problems, high blood pressure, and heart disease.

Yehuda wasn't sure what she would find in the urine in her parents' freezer. "I had no clue," she insists, but she was pretty sure the Holocaust survivors were different from the vets. When she measured their stress hormones, however, they were actually quite similar. Many Holocaust survivors had unusually low cortisol levels. Fifty years after their greatest traumas, they were still vulnerable to PTSD or were suffering from it. Yehuda's interviews with the survivors revealed the extent of their trauma. In one conversation, she asked a woman in her nineties: "Do you have intrusive thoughts" about the Holocaust? "Of course," the woman answered. "The Holocaust is with me all the time." She woke up at least once a week with nightmares of Nazi guards chasing her with German shepherds. Typically, the dream was triggered by a dog barking somewhere in the neighborhood. Usually, she was able to console herself and go back to sleep. "I cry, thanks God," the woman said in the grammar and accent of a Czech immigrant. "I can wake up! Once it was my life. I couldn't wake up."

In subsequent research, Yehuda found that the psychological shadow of the Holocaust loomed large even in the offspring of the survivors. Indeed, she discovered that the children of survivors also produced lower cortisol levels and were more vulnerable to trauma and stress reactions including depression, anxiety, and PTSD. It's a mysterious phenomenon labeled intergenerational syndrome, and it wasn't limited to the Holocaust. Yehuda found that pregnant women traumatized by the September 11 attacks on America may have passed on stress markers to their unborn babies in a ripple effect. Studying thirty-eight pregnant women who suffered from anxiety and other PTSD symptoms after 9/11, she found that they produced lower levels of cortisol—and their babies did, too, when they were measured around their first birthdays. Somehow this marker for stress vulnerability had been passed from mother to newborn. In the children of Holocaust survivors, Yehuda believed these lower levels were caused by the stress of living in close proximity to anxious or depressed parents and hearing frequently about the traumas of war. But the 9/11 babies were different. They couldn't have known about the destruction of the World Trade Center. "There is some biological transference from the mother to the fetus," Yehuda says. The transgenerational transmission of trauma may be caused by "very early parent–child attachments," she guesses, or possibly the effects of cortisol programming in the womb.

Today Yehuda works in Building 107 of the huge, gray VA Medical Center in the Bronx. She came here in 1991 to start a PTSD clinic, which now helps around six hundred veterans with the ongoing trauma of war. Under the fluorescent lights of a drab government building, she seems unusually fashionable in sleek glasses, an enveloping black shawl, and silver running shoes. She leads me on a quick walk through the rooms where vets come to hang out, drink coffee, lift weights, and swap stories. At age forty-nine Yehuda is a leading authority on PTSD and resilience and has just returned from an international conference in Santiago, Chile. She exudes warmth, caring, and curiosity and has devoted her career to helping heal the psychological wounds of war. She is also feisty, passionate, and outspoken, a poker player who likes to take risks and eviscerate flabby thinking. Above all, she wants to demolish the idea that resilience is a constant, steady state. Instead, she believes the experience of Holocaust survivors shows that people can be different things at once. It may seem obvious, but Yehuda says many experts—and the media—want to believe that people are either resilient *or* vulnerable, strong *or* weak, healthy *or* sick. In reality, people combine all these qualities. Holocaust survivors are remarkable and successful but they also suffer.

After our meeting in Building 107, Yehuda reached out with another story to illustrate her point about resilience. Sixteen years ago, she met a remarkable survivor. To protect her identity, Yehuda calls her Magda—not her real name—and changes a few other details about her life. During the war, Magda and her father hid in a forest in Poland. Her father told her it was vital to stay alert, especially during the night. One morning, Magda woke up horrified to discover that she had fallen asleep. Even more terrifying, her father was gone. Not long after, Magda was caught by the Nazis and dispatched to Auschwitz.

Magda survived her imprisonment, and after the war she married another survivor and immigrated to an affluent suburb in Westchester, New York. The couple worked very hard and prospered. They had two children who grew up to be accomplished professionals. Over the years, Magda experienced some PTSD symptoms and episodes of major depression, but she functioned remarkably well. In interviews with Yehuda, Magda spoke a lot about the coping skills that allowed her to survive and thrive, including taking pleasure in the achievements of her husband and children. She felt survivors had an obligation to live their lives fully and to make the most of every moment. She believed that success and happiness were important posthumous victories over Hitler. When she talked about her

Holocaust experiences, she sobbed. But each time, she pulled herself together.

Some ten years after their first interview, Yehuda went back to see Magda. The news wasn't good. In the previous two years, she had lost her son and grandson to cancer. Her husband had also died. Her family was extremely concerned about her depression and negativity, but she refused any treatment or medication. Magda's daughter asked Yehuda to convince her mother to see a psychiatrist so she could get a prescription for antidepressants. But Magda refused. She deserved to suffer because she couldn't protect her loved ones from death. She had survived the Holocaust in Europe but was unable to save her own son and grandson in America. She only had herself to blame for believing there had been a purpose for her survival in the concentration camp. Surely, God was not only mocking her but punishing her, too. First, He had taken her family during the war. Then He kept her alive so that she would see her son and grandson die. Now He wanted her to watch her fourteen-year-old granddaughter cope with losing her father and only brother. If God intended for her to suffer, why numb the pain with medicine and deny her fate?

At first, Yehuda didn't respond. She felt paralyzed by Magda's anguish. For a long time, neither spoke. Finally, Yehuda said: "I don't know why God does what He does, and I am very sorry to see how much pain you are in. But I can't help thinking that under the circumstances, your granddaughter is pretty fortunate."

"Fortunate?" Magda replied angrily, as if to say, *Are you mocking me, too?*

"Fortunate," Yehuda said, "to have someone in her life who can understand what she is going through. You of all people know what it is like to lose a father and brother at this age. You were not so lucky to be able to cry in the arms of your grandmother, but I imagine that it would have been much easier for you if you had. Imagine if even one person from your family had been there so that you would not have been all alone. Your granddaughter has you. Maybe that's why your daughter wants you to take those antidepressants, since you cannot be of much help in your current condition. No one would blame you if you have reached your pain threshold and can't pick yourself up again, but that doesn't mean there is no reason to."

Yehuda's goal was to remind Magda that she was only being asked once again to survive. "Undoubtedly," Yehuda says, "what had helped her survive before—her commitment to her descendants—might help her in exactly the same manner." She continues, "If we think of resilience as a stable trait, we run the risk of becoming paralyzed or confused if for a moment it seems to disappear. If we think of resilience like we think about the sun, then when we can't see it, it just means we have to wait a few hours for its return. But perhaps even this is too passive a metaphor. Maybe it is better to think of resilience as a muscle to build, maintain, and remind oneself or others to flex."

Each one of us can play "a role in building and maintaining this muscle," Yehuda says. "We should not talk about resilience as some capacity that you are born with. Though it is easier for some people to build and maintain this muscle, it is ultimately and optimally developed and maintained through constant work." Today Magda is doing well. "She's a tough lady," Yehuda says. She's frailer now and her resilience muscles are older, but she's very much a survivor. She knows how to suffer *and* how to recover and rebuild.

Yehuda leads me through crowded hallways of the VA. Our walk takes us past men in wheelchairs and others with canes. Look around, she says, gesturing at the swarm of people who need medical care. "What keeps people going is the dialectic between their strong parts and their broken parts. They wouldn't be here trying to get well if they didn't believe in their own resilience." We keep going, and Yehuda explains that the world isn't divided into resilient people and nonresilient people. Handling adversity is a seesaw struggle. The goal is to find a "steady state" in life where we can feel good more than bad, strong more than weak. The Holocaust survivors aren't necessarily better or more resilient or wiser than the rest of us. They're just more challenged. And with greater adversity comes more opportunity for recovery and resilience.

"We're all made of the same DNA," Yehuda says. We're all fighting to stay alive, to bounce back from hardship, to make something of our precious time here. And trauma is inevitable. So we have to prepare ourselves and our children. Ever since September 11, Yehuda's twelve-year-old daughter Rebecca sometimes admits how scared she is that bad things might happen. "I'm always tempted to reassure her that harm will not come her way," Yehuda says. "But then I think of all the Holocaust survivors and I cannot promise this. Instead I reassure her that she will have the strength and resourcefulness to cope with whatever challenges her." Then they practice what to do in a crisis so that her daughter can feel more ready. "I think every single one of us has to be armed and ready to confront trauma," Yehuda tells me. "The day will come." If we face this fact, we can equip ourselves by saying: *Of course this will happen to me and this is what I will do.*

8

The Science of Luck

Why Good Things Always Happen to the Same People

Professor Richard Wiseman can tell if you're lucky or unlucky just by handing you a newspaper and asking you to count the number of photographs in its pages. Some folks finish the job in a few seconds while others need a couple of minutes to tally all the pictures. The reason for the difference isn't that some people are better counters than others. Rather, the secret lies on page two of the newspaper where Professor Wiseman has inserted a huge message in inch-and-a-half letters that are even larger than these:

STOP COUNTING—
THERE ARE 43
PHOTOGRAPHS IN
THIS NEWSPAPER

Believe it or not, many people actually miss this enormous headline in the paper. They're too busy counting photos to notice. The giant message isn't a trick. There are really forty-three pictures in the paper. Professor Wiseman has found that if you see the announcement right away, you tend to be a lucky person open to random opportunities around you. By contrast, if you don't spot it, you're usually an unlucky person more likely to miss out on fortuitous possibilities. But the experiment doesn't stop here. Wiseman inserts one other message halfway through the newspaper in giant letters:

STOP COUNTING,
TELL THE
EXPERIMENTER
YOU HAVE SEEN
THIS AND WIN $250

Once more, the unlucky people are too focused on adding up photos. They don't even see the message and lose out on the chance to win $250. By comparison, lucky people often spot the opportunity and claim the prize. Even when they're doing one job, they're open to other opportunities. This is one of the big reasons Professor Wiseman believes good things always happen to the same people. In everyday life and extreme survival situations, this is frequently what we call luck.

* * *

Psychologists use the phrase *inattentional blindness* to describe what Professor Wiseman captured in his newspaper test. It's a polysyllabic way of saying that we don't notice things when we don't pay real attention. One of the most famous studies of inattentional blindness was conducted by Daniel Simons and Christopher Chabris in the elevator lobby of the fifteenth floor of the Harvard Psychology Department. One team of players wearing white shirts and another group dressed in black tossed two orange basketballs back and forth. Subjects were asked to watch a video of this ball-passing exercise and count the number of passes made by players dressed in white. After forty-five seconds in one version of the video, a woman in a full gorilla costume walks right through the scene. The hairy ape is clearly visible crossing the screen for five seconds. Remarkably, 56 percent didn't even notice the gorilla right in the middle of the action.* In another video, the gorilla stops, faces the camera, pounds her chest, and then marches off. The action lasts nine seconds, but again only 50 percent spotted the furry interloper.

The "Gorillas in Our Midst" inattentional blindness experiment.

* To see the video, go to http://viscog.beckman.uiuc.edu/media/survive.html. It is also available as part of a DVD from Viscog Productions at www.viscog.com.

How is it possible to miss the gorilla? And what does it tell us about survival? Professor Simons now teaches psychology at the University of Illinois at Urbana-Champaign. The main lesson and surprise of the gorilla experiment, he tells me, is how easy it is to miss something as obvious as a gorilla. "Distinctive and unusual objects do not automatically capture our attention," he says. Many other studies have demonstrated that it's difficult if not impossible to be aware of everything going on around you—or even right in front of you. One reason is that your eyes only see in high resolution within two degrees of your focal point. In other words, no matter how good your eyesight, the majority of your surroundings are essentially out of focus. Try holding your arm out in front of you and making the thumbs-up sign. The sliver of the world that you see in high resolution is only about as wide as your thumbnail. If you focus, say, on your cuticle, you'll immediately notice how the detail in your peripheral vision drops off dramatically. Now try staring at the letter *c* in the middle of the word AVOCADO. Can you also identify the words just two to the right or left? You probably can't. That's why, when you read this sentence or cross the street, your eyes move three or four times every second to absorb information and create the impression in your brain of seeing everything in high resolution. In fact, you're glimpsing only a tiny slice of reality at any given moment.

The gorilla experiment is important, Simons says, because it shocks you into realizing how little of your environment you consciously perceive, especially if you're very focused on a specific task. Once you've gained this insight, you can start opening yourself up to all the possibilities that you may be missing. For instance, whenever he drives his Honda Accord, Simons tries to anticipate unexpected events because he knows they won't *automatically* capture his attention. Specifically, he's on guard for "things that are most likely to be catastrophic." When he crosses a traffic intersection, he always checks for red-light runners because he's familiar with accident reports that quote drivers saying, "I was looking but I didn't see the other car," or "The vehicle that rammed me came out of nowhere." In reality, the cars were always there but the driver simply didn't *see* them. In everyday life, Simons recognizes there's no guarantee he'll notice a gorilla or cement truck coming right at him. This awareness has changed the way he interacts with the world.

When it comes to spotting hairy apes and red-light runners, Professor Wiseman believes there is also another important factor at work. Neuroticism is a personality trait of people who tend to be anxious, tense, and sensitive to stress, he explains. In the gorilla experiment, people with high levels of neuroticism are very serious and intense about their assignment to count the number of basketball passes. People with low levels are more calm, relaxed, even-tempered, and emotionally stable, and less sensitive to stress. They carry out their assignment but aren't as anxious or intense. According to Wiseman, lucky people usually have lower levels of neuroticism. They're more laid-back and open to life's possibilities—like giant headlines in his newspaper experiment offering cash prizes—while unlucky people are more uptight, nervous, and closed off.

If you want to test yourself, take a quick look at this domain name:

WWW.OPPORTUNITYISNOWHERE.COM

What do you see? For many people, the Web site seems discouraging: Opportunity is *nowhere*. But others see the exact opposite: Opportunity is *now here*. When it comes to hidden messages, lucky people perceive more of the world around them. "It is not that they expect to find certain opportunities, but rather that they notice them when they come across them," Wiseman writes in his fascinating book *The Luck Factor*. This ability or talent "has a significant, and positive, effect on their lives."

* * *

As you may have guessed from his newspaper experiment, Richard Wiseman is not your typical academic. Bald, bespectacled, goateed, and forty-something, he works at the University of Hertfordshire in the south of England and holds Britain's only professorship in the public understanding of psychology. In his teens, Wiseman fell in love with magic, joined London's renowned Magic Circle, and performed at Hollywood's legendary Magic Castle. Endlessly curious about the quirks of human behavior, he has written two best-selling books, published forty papers in leading journals, and made headlines around the world for his surprising quests to discover the world's funniest joke and the best pickup line.* Devoting a decade to exploring the secrets of rabbit feet and four-leaf clovers, Wiseman discovered that some people actually do have all the luck while others are a "magnet for ill fortune." Naturally, he wondered: Are some people born lucky while others are star-crossed? The answer is no. "There were too many people consistently experiencing good and bad luck for it all to be chance," he writes. "Instead, there must be something *causing* things to work out consistently well for some people and consistently badly for others."

Most people define *luck* as "an unpredictable phenomenon that leads to good or bad outcomes in life." In this view, the keys are unpredictability and randomness. Luck is something that happens by pure chance. If you survive an airplane crash in seat 10A while someone in 10B perishes, that's pure luck. As he does with so many topics, Wiseman turns this traditional

* The funniest joke in the world goes like this: Two hunters from New Jersey are in the woods. One collapses and the other phones for help, telling the operator he thinks his friend is dead. When she asks if he's sure, there's a gunshot before he comes back on the line to say, "Okay, now what?" Wiseman doesn't think the joke is especially funny, but it won an online competition. The best pickup line in the world: "What's your favorite pizza topping?" Wiseman explains that it's an open-ended question that generates a lot of conversation possibilities.

thinking upside down. The passenger in Seat 10A may have survived by chance, but Wiseman believes if you look more closely, there may be more to the story than dumb luck. For instance, the survivor's attitude and actions may have helped increase her odds of getting out alive. She may have paid careful attention to the safety briefing, memorized an escape plan and evacuated immediately without waiting for the flight attendants to tell her what to do. "Luck is not a magical ability or a gift from the gods," Wiseman writes. "Instead, it is a state of mind—a way of thinking and behaving." Above all, he insists that we have far more control over our lives—and our luck—than we realize. Since the Italian Renaissance great thinkers and writers have argued that 50 percent or more of what happens in life is determined entirely by chance or Fortuna, the Roman goddess. Wiseman says no way. He believes that only 10 percent of life is purely random. The remaining 90 percent is "actually defined by the way you think." In other words, your attitude and behavior determine nine-tenths of what happens in your life.

If you're worried that you would have skipped over the giant messages in the newspaper experiment, fear not. You may feel unlucky, but Wiseman says that within weeks, you can learn the secrets of his Luck School and increase your good fortune by more than 40 percent. It may seem improbable or even impossible, but he says it's scientifically proven. You'll probably find more money on the ground. You'll improve your chances of meeting your soul mate. You'll even boost the likelihood of clinching the business deal that's eluded you for years. And in terms of joining the Survivors Club, you may even save your life.

1. The Luck Formula

What is luck? Brace yourself and take a gander at this equation:

$$\lambda(E) = \Delta(E) \times [1\text{-pr}(E)] = \Delta(E) \times \text{pr}(\text{not-}E)$$

It's all Greek to me, but not to Nicholas Rescher of the University of Pittsburgh who has written more than eighty books on philosophy, including an engaging volume on luck. He uses this equation to calculate the amount of luck involved in any given event. It's too elaborate to explain in these pages—frankly I could never do it justice—but Rescher tells me the formula boils down to two main considerations: the stakes and the probability of success. The bigger the stakes and more unlikely the success, the luckier you are. Conversely, the smaller the stakes and the more likely the success, the less lucky you are. In short, Rescher writes, *luck* is defined as "good or bad fortune acquired unwittingly, by accident or chance." He continues: "We are, all of us, at the mercy of unpredictable developments that make it a matter of mere luck how many of the crucial issues in our lives are resolved. The fact is that most human enterprises are to some extent chancy. Effort alone is seldom sufficient because of all the things that could go wrong: a mixture of effort and luck is often needed to achieve success."

If that's true—if a combination of action and luck is necessary for success—doesn't that formula also hold true for survival? The answer is yes. Many survivors told me that lucky breaks played a significant role in overcoming their adversity. But this begs another question: Do some people get more lucky chances than others? Again, the answer is yes. To understand the reasons, Richard Wiseman ventured to the center of London to interview regular people. Fifty percent claimed they were lucky in their lives, while 14 percent claimed to be unlucky. The rest—36 percent—believed they were neither. Wiseman wasn't satisfied. He wanted to figure out the differences between lucky and unlucky groups. Over the course of a decade, he tested lucky and unlucky people in the most ingenious ways and concluded that there are four reasons why good things always happen to the same people.

First, lucky people constantly happen upon chance opportunities. "Being in the right place at the right time is actually all about being in the right state of mind," Wiseman writes. As his newspaper experiment shows, lucky people are more open and receptive to unexpected possibilities. They tend to be more relaxed about life, and they operate with a heightened awareness of the world around them. Quite simply, they spot and seize upon openings that other people miss. They also tend to be more social and maintain what Wiseman calls a "network of luck." Most of us know around three hundred people on a first-name basis. Wiseman says that means you're only two introductions away from ninety thousand people who could bring chance opportunities into your life. If you invited fifty people over to dinner, that means you're only two degrees of separation from four and a half million potential lucky breaks.

Second, lucky people listen to their hunches and make good decisions without really knowing why. Unlucky people, by contrast, tend to make unsuccessful decisions and trust the wrong people. "My interviews suggested that lucky people's gut feelings and hunches tended to pay off time and time again," Wiseman writes. "In contrast, unlucky people often ignore their intuition and regret their decision." In survival, this kind of instinct can make all the difference. Wiseman describes a twenty-four-year-old dancer named Eleanor from California who was driving to her parents' home one night when she noticed a motorcyclist behind her. She assumed he was lost, but when she arrived at her destination, the biker pulled alongside her car. Eleanor suddenly felt very cold and was overcome with a strange sensation. "I can't explain it but I knew he had a gun and wanted to kill," she told Wiseman. So she stayed in her car, started the ignition and got ready to speed away. The cyclist appeared to get nervous and took off. Eleanor immediately called the police. Two days later, an officer stopped the same motorcyclist. This time, the biker pulled a gun and murdered the cop. Wiseman writes: "Eleanor is convinced that her intuitive decision to start her car saved her life." Over and over, survivors recount how gut feelings and intuition helped them overcome adversity. Their instincts are "often surprisingly accurate," Wiseman writes. "And even more amazingly, they have no idea what lies behind their success. To them it just looks like luck. In reality, it is all due to the remarkable inner workings of their unconscious minds."

Third, lucky people persevere in the face of failure and have an uncanny knack of making their wishes come true. Perhaps it's just plain old-fashioned optimism—seeing the glass half full—but lucky people expect good things to happen. They're convinced that life's most unpredictable events will "consistently work out for them." They keep going in the face of failure until they accomplish their objectives. Their world is "bright and rosy," Wiseman writes, while unlucky people *expect* that things will always go wrong. Their world is "bleak and black." When Wiseman gives lucky and unlucky people a puzzle that is actually *impossible* to solve, the reactions are very telling. "More than 60 percent of unlucky people said that they thought the puzzle was impossible, compared to just 30 percent of lucky people. As in so many areas of their lives, the unlucky people gave up before they even started."

Fourth, lucky people have a special ability to turn bad luck into good fortune. Of all four defining factors involved in luck, Wiseman believes this one plays the most important role in survival. Wiseman's conclusion echoes the work of Dr. Al Siebert, one of America's foremost authorities on survival psychology. After more than forty years investigating what he calls the survivor personality, he believes that one of the most critical skills is what he terms the serendipity talent. "When misfortune strikes," Siebert explains, "life's best survivors not only cope well, they often turn potential disaster into a lucky development." He cites the eighteenth-century English writer Horace Walpole, who coined the word *serendipity.* It comes from the Persian fairy tale about the three princes of Serendip who were banished by their father, King Giaffer, so that they could learn practical lessons about the world. On their journey, the three princes used their cleverness to turn hardship into good fortune. Their resourcefulness and a series of fortuitous events led them to marriage, wealth, and their own kingdoms. "Serendipity is not good luck," Siebert insists. Rather, it's a skill that can be developed. This process of turning adversity into advantage happens with "amazing speed," Siebert writes, and it's so important in survival that he considered calling his book *The Serendipity Personality* instead of *The Survivor Personality.* "In my mind," he explains, "they are interchangeable.

"Life's best survivors react to disruptive change forced on them as though it is a change they desired," he continues. "The same crises or disruptive changes that make some people victims are turned into good fortune by people with the serendipity talent."

2. Why Bronze Medals Are Better Than Silver

When you hear her story, you might think Cindy Roper is one of the unluckiest people on earth. In the summer of 2000 in the foothills near Alamogordo, New Mexico, her horse Frenchy slipped on a country road, fell onto the blacktop, and kicked her in the chest at close range. A bunch of Cindy's ribs broke, and one lung collapsed. In the ambulance, paramedics jabbed a large-bore needle into her chest to relieve the pressure. They told her if she survived, she should celebrate this day as her second birthday. And so on August 10 every year, she lights candles to remember her close call. "I'm glad I'm alive," she tells me. But that was hardly the end of Cindy's bad luck. On Memorial Day 2007, she came down with a high fever and headache that wouldn't go away. She lives with two dogs on a remote ten-acre property in the hills of the Ortiz Mountains around thirty miles from Santa Fe. On the fourth day of burning fever and throbbing pain, Cindy could barely take care of herself, so she called an ambulance. Paramedics measured her temperature: It was 104 degrees. At St. Vincent Hospital in Santa Fe, doctors were baffled. Initially, they believed she was suffering from spinal meningitis. But a few days later they rushed into her hospital room with a stunning diagnosis.

"It's looking plague-like," they said.

Cindy couldn't even wrap her mind around the word *plague*. Every year in the United States, ten to fifteen people come down with the plague, a fatal disease if untreated. Around the world, around two thousand cases are reported, 98 percent of them in Africa. The disease is carried by rodents and transmitted to humans by fleas. Cindy was diagnosed with what's called septicemic plague, because the bacteria multiply in the blood. Investigators scoured her property searching for the dreaded disease and determined that she was probably infected by a flea from a rock squirrel that lived under a bush on the path to her clothesline. Within days, the antibiotics worked their magic and Cindy felt fine again.

If you were nearly killed by your horse or infected with the plague, you might think you were cursed or jinxed. But not Cindy. "I feel lucky," she tells me at breakfast near the historic plaza in Santa Fe. She's fifty-two years old and bears a resemblance to the actress Jamie Lee Curtis, with lively brown eyes and short spiky hair graying on the sides. She wears a salmon-colored shirt and a mother-of-pearl pendant in the shape of a bear. It's a Native American carving by Zuni artists, and Cindy believes it gives her energy and healing.

"I come from very strong women," Cindy tells me. "I'm very self-sufficient." For years, she worked as a police dispatcher answering 911 calls. Every day, she talked to people in the worst moments of their lives. Sometimes she heard gunshots over the phone or the screams of domestic violence. She knows terrible things happen all the time. She looks around at the other diners sipping coffee and eating breakfast burritos. "We're all survivors in one way or another," she says. "I don't feel that I'm stronger than any person you see today."

Sure, she's come close to death by horse. Yes, she contracted the plague. But she doesn't feel unlucky. Her eyes well up with tears as she runs through the litany of her good fortune. She's got a beautiful daughter. A wonderful network of friends. Animals she adores and that adore her. "How can anyone perceive that as unlucky?" she asks. "I don't see it as unlucky."

Lucky people like Cindy always see the positive side of their misfortunes. Psychologists call this ability counterfactual thinking. In other words, lucky people can readily imagine how much worse off they might be. Cindy, for instance, is well aware that she could easily have died twice. The same concept holds true for Olympic athletes. Psychologists at Cornell University discovered that bronze medalists were actually happier than silver medalists. At the 1992 Summer Games in Barcelona, it turns out that silver medalists were consumed with thoughts of "nearly winning the gold." They framed their glory not in the triumph over everyone else at the Games but in their loss to one person. Whatever elation they might have felt, the researchers found, "is often tempered by [torturous] thoughts of what might have been." On the other hand, when bronze medalists engaged in counterfactual thinking, they realized that they might have finished in fourth place and ended up "in the showers instead of on the medal stand."

Richard Wiseman has his own way of testing luck and counterfactual thinking. It goes like this: How would you feel—lucky or unlucky—if you were standing in line at the bank when an armed robber entered, fired one shot, and hit you in the arm? Take a moment and think about it. You're cashing your paycheck and you're wounded in the arm by a bank robber. Are you lucky or unlucky? Wiseman found that lucky people tend to think that the situation could have been far worse. In their counterfactual thinking, the bullet could have killed them. Unlucky people see the hypothetical very differently. A bullet in the arm is bad luck, plain and simple. In another scenario, Wiseman asks you to imagine slipping on loose stair carpet, falling down a flight of stairs, and twisting your ankle. Lucky or unlucky? Again, thinking counterfactually, lucky people believe they're fortunate that they didn't break their necks. Unlucky people believe that falling down the stairs is bad news no matter what happens.

"The differences between the lucky and unlucky people were amazing," Wiseman writes. "Many unlucky people saw nothing but misery and despair when they imagined themselves experiencing the bad luck described in the imaginary scenarios. Lucky people were the opposite. They consistently looked on the bright side of each situation and spontaneously imagined how things could have been worse. This made them feel better and helped maintain the notion that they were lucky people living lucky lives."

3. Why Accidents Never Really Happen

If some people have all the good luck, does it follow that other people have all the bad luck? If some folks never break their bones or crash their cars, do others always end up in the emergency room? To answer these questions, I traveled to the town plaza in the city of Orange, California. It's a sunny day, traffic spins around the oval in the heart of the historic district, and the sidewalk café is crowded. I'm a little wary about the woman I'm supposed to meet. Given her calamitous history, I worry about sitting with her at a curbside table so close to moving vehicles. But Samantha Dunn doesn't seem like an accident waiting to happen. She looks healthy and wholesome. Her hair is auburn, her eyes are blue, and she's got a winning, boisterous personality that she attributes to her Irish Catholic genes. Sam used to call herself Calamity Jane, but her mother put it differently. "Some people are born rich," her mom would tell her. "You were born a klutz." Young Sam was always splitting her lip, squashing her finger in a car door, bumping her head, or twisting her ankle. "I was the kid who was going to get hurt," she tells me. Her litany of injuries reads like an emergency room log: broken wrist, toes, and rib; multiple concussions; second-degree burn from a motorcycle exhaust pipe; too many sprains to count; cracked coccyx; torn rotator cuff; multiple scars across the knuckles.

Sam recounts each tale of woe with dramatic flair and self-effacing humor. Her worst accident came on the afternoon of September 15, 1997. At the age of thirty-two, she was a health and fitness editor and journalist working for various national magazines. One day on a Malibu canyon trail in Southern California, Sam was leading her horse Harley through a creek when he was startled by running water. The two-thousand-pound animal reared up and trampled her. Harley's steel-shod hind hoof cut straight through Sam's left shin "like an ax chopping a tree." The sound of the shattering bone was "loud as gunfire." Sam was unable to walk or crawl. With each heartbeat, she watched her own blood spurt into the dirt "like water from a garden hose left running on the lawn." Sam could hear her own breath and smell the earth and sweet grass around her. A scrub jay was making a racket, and the wind was ruffling the oak trees. *People have died to these sounds for thousands of years,* she thought. *I can't die this way today.* So she screamed for her life. Suddenly a dark-haired, green-eyed man appeared. Despite her trauma, she vaguely recognized him as an actor from the movie *Butterflies Are Free* with Goldie Hawn. His name was Edward Laurence Albert, son of Eddie Albert of *Green Acres* and *Brother Rat* fame. He touched her left shoulder and said in a deep voice, "I'm trying to remember anything useful from the times I've played a doctor."

"I feel better already," Sam deadpanned.

He phoned 911 with one hand, and then pinched an artery in her leg to stop the bleeding. He also stroked her hair. While waiting for paramedics, he tried to keep her focused and alert. "The way two people talk when one is keeping the other from bleeding to death makes for its own category of conversation," Sam writes in her funny, fascinating memoir, *Not By Accident*. When help finally arrived, she was rushed by helicopter to UCLA's trauma center.

As she healed from what doctors deemed a serious leg fracture, Sam began to wonder why her life seemed like an endless series of mishaps. Was she really a klutz? Was she jinxed with bad luck? Or was there something else going on? "I am blue-eyed because of genetics; I am math-phobic because of bad teaching; I am redheaded because my hair-dresser stands behind my refusal to accept brown hair," she writes. But she couldn't pinpoint the underlying reason for her accidents. Using her journalism skills, she began to call experts across the country. Her search took her back to Sigmund Freud's 1899 book *The Psychopathology of Everyday Life* and his theory that accidents were "expressions of unconscious intent, and the more severe the accident, the more clearly it expressed a desire for self-destruction." She interviewed academics who studied why people with marital problems get into more car wrecks and why children undergoing stressful life changes are more susceptible to accidents. She also began to unpack her own life. She inquired whether being left-handed in a right-handed world was the reason for her many woes, a controversial idea treated in depth on page 205. She explored whether her problems might stem from an undetected brain tumor or early-onset dementia, but doctors dismissed both possibilities. After ruling out the more obvious explanations, she found studies suggesting that people who experience repeated accidents often suffer from psychological disorders like depression and anxiety. "If you've hurt yourself once," she writes, "it's likely that you will hurt yourself again."

At the end of her investigation, Sam believes she discovered the reasons for her "life of injury." She writes: "I think, somewhere along the line, I figured out that letting 'the world' hurt me served a few functions extremely well: It provided me with a kind of nurturing I didn't otherwise know how to attract, I couldn't be pinned with total responsibility, and it provided physical pain, a reason to cry that others could understand. So much easier than trying to explain all the accumulated rage and numbness and sadness."

Dr. Ellen Visser knows accidents and countless survivors like Sam Dunn. In 2007, the Dutch medical researcher gathered information on 147,000 people who'd experienced accidents and injuries around the world. After crunching all the numbers, she concluded that despite the very definition of an accident—an unfortunate event that occurs unexpectedly and unintentionally—they're almost never accidental. "Accidents are not distributed by chance," she tells me from her office in Groningen, the largest city in the north of the Netherlands. Certain people seem to be at higher risk of repeated accidents. Specifically, she found that people in this accident repeater group were 1.42 times more likely to have an accident than chance alone would have dictated. Simply put, accidents tend to cluster around certain people. After reviewing seventy-nine conflicting studies on the subject, she concluded: "Accident proneness exists." The people who are most at risk tend to be more hyperactive, impulsive, neurotic, extroverted, and—no surprise—inclined to use alcohol or drugs. Yes, people encounter bad luck, Visser believes. But people are largely responsible for their misfortune. Her latest research project at the University Medical Hospital in Groningen is focused on injury repeaters in the emergency room. Typically, they're younger men who have more self-destructive tendencies than the rest of us. At a *conscious* level, they have more suicidal ideas, plans, and attempts. At an *unconscious* level, they pursue unhealthy lifestyles of smoking, obesity, and physical inactivity.

If you carefully investigate any accident, experts say, it probably isn't an accident after all. As sociologist David Phillips says, "To use the word *accident* is to admit ignorance about the cause of something." So a car crash may look like an accident, and that's the way the highway patrol officially reports it. But in fact, there may have been all sorts of underlying reasons that weren't really accidental. Indeed, the crash may have happened "accidentally on purpose," as Ellen Visser puts it. In fact, she explains, some car crashes might actually qualify as an indirect form of suicide. You're depressed. You're angry. You're not happy with your life. You neglect to take care of yourself and your car. You drive a little too fast. You make a reckless turn that's too sharp. Wham. Your tire blows out. Is it really an accident? Or was it a death wish?

In 2001, the prestigious *British Medical Journal* banned the word *accident* from its pages. The editors argued that accidents are mistakenly believed to be unpredictable, chance occurrences and "one of the most common and familiar forms of bad luck." In fact, the editors reasoned, "most injuries and their precipitating events are predictable and preventable." They acknowledged that some injuries from earthquakes, avalanches, lighting strikes, and the like are real accidents that can be attributed to bad luck or acts of God. But even those cases are debatable as accidents because "with modern technology it is often possible to predict when these events will occur." The journal even suggested a new word to describe incidents that produce injuries: *injidents*.

For years, accidents—or *injidents*—materialized in Sam Dunn's life like some kind of bad rash, but it's been a long time since she's experienced a flare-up. "I just make different choices now," she says. As she stirs an iced latte, she recounts her recent run of good luck. Despite her liberal leanings, she fell in love with a Republican, got married, and gave birth to a baby boy. She lives in a charming old house, she rides horses, and she's writing a new book. On the third finger of her right hand, she's wearing a silver Italian spoon ring that was given to her by a woman she met on a plane flight to New York. The passenger's name was Darla, and just like Sam, she was a tall redhead who was recovering from a serious leg injury. She was around fifteen years older and she popped drinks and painkillers all the way across the country. Darla was a "train wreck," Sam says, but in a strange way she was also very recognizable. Indeed, Sam believed Darla was her mirror image fifteen years in the future. At the end of the trip, they exchanged rings. Sam pulls the silver band from her finger. She's pretty sure that Darla isn't alive today because she was so unhealthy and self-destructive. Sam wears the ring every day. "It reminds me to take care of myself," she says, because it would be so easy to make different choices that would inevitably lead to more accidents and ultimately death.

Ninety percent of your life and your survival come down to the small choices you make, Sam argues, echoing Richard Wiseman. It seems almost too mundane to mention, but did you check the air pressure in your tires? Did you replace the batteries in the smoke detector? Did you bother to check your ski bindings? Did you tighten the straps on your child's car seat?

Did you get your annual mammogram or regular Pap smear? Are you in denial about your partner's addiction problems? Yes, bad luck strikes and awful things happen. But nine out of ten times, we bring misfortune and accidents upon ourselves, she believes. If you want to break the cycle, you've got to change your mind-set and take responsibility. Who gets into the Survivors Club? "The people who make the choice to survive," Sam insists.

"The truth is that I am not prone to accidents," she writes in her memoir. "I am prone to jumping on the back of a horse and riding until the rush of wind in my ears erases the screaming and name-calling between the two women I love most in the world. I am prone to going to the stable rather than asking my (ex) husband if he can please take that . . . bong out of his mouth. I am prone to panic over when I'm going to get paid for my last freelance article and whether the check will arrive before the phone bill is due. I am prone to fly off the handle. I am prone to be sad for reasons that I don't really understand. I am prone to want to see myself bleed rather than admit I'm scared. Accidents are just part of the deal."*

"'Accident-prone' is not a description of my character," she has written. "It's a state of mind I enter." Survival means mindfulness. To live well and dodge bad luck, you need to focus on the "state of mind of good possibilities," Sam tells me, "or the state of mind of safer possibilities."

4. The Secrets of Luck School

Richard Wiseman wasn't satisfied simply studying luck. He wanted to see if he could actually help make it. So he created what he calls Luck School to put his ideas to work. It's not a

* Sam divorced her first husband and is happily remarried.

real academy with classrooms, blackboards, and final exams. Rather, it's a five-step program that he's tested with great success on regular people. As outlined in *The Luck Factor,* the first stage involves signing a "luck declaration" in which you pledge to incorporate his principles into your life for one month. This oath signifies you're prepared to make a real investment and necessary changes to turn your luck around. Second, you create your luck profile by filling out a short questionnaire. The goal is to identify your strengths and weaknesses when it comes to fortune. Do you maximize chance opportunities in life? Do you listen to your lucky hunches? Do you expect good things to happen? Do you turn bad luck into good? The third step involves incorporating some of Wiseman's techniques into your daily life. For instance, you're supposed to expand your network of luck and meet new people; practice a more relaxed attitude to life; open yourself to new experiences; and pay more attention to your gut instincts. The fourth step involves keeping a luck journal and jotting down fortunate events that happen to you. Finally, Wiseman urges people to take their time creating a lucky life. "Good fortune experienced by lucky people," he writes, "is not the result of the gods smiling on them, or their being born lucky. Instead, without realizing it, lucky people have developed ways of thinking that make them especially happy, successful and satisfied with their lives." From the outside, it may look like they're lucky and their lives are charmed. "Deep down," Wiseman writes, "they are just like you . . . and you can be just like them."

The results of this five-step Luck School are impressive even for people who don't feel especially fortunate in life. Wiseman says that 80 percent of the participants in the program report that their luck improves at least a little. The average increase is more than 40 percent, and it takes only a month. He's also tried out Luck School in the business world. One computer company in Hertfordshire, England, put around forty employees through the course. Every month, productivity jumped 20 percent, and the managing director was delighted. The same principles, Wiseman believes, apply to survivors, too. "When it comes to luck," he writes, "the future is in your hands."

Survival Secrets

Where Have All the Lefties Gone?

Nine out of ten people in the world are right-handed. But here's a surprise: The older you get, the fewer lefties you'll find in your age bracket. Among ten-year-olds, 15 percent are left-handed. Fast-forward to fifty-year-olds and the lefty share shrinks to only 5 percent. Jump to eighty-year-olds and—remarkably—fewer than 1 percent are left-handed. So where do all the lefties go? And is some sinister* force involved?

For years, Dr. Stanley Coren investigated what he calls "the case of the disappearing southpaws." Mysteriously, righties outnumber lefties two hundred to one by the age of eighty-five, says Coren, a right-handed professor of psychology at the University of British Columbia. How is this possible? One explanation—the so-called modification theory—is that lefties turn into righties over time. To fit into the right-handed world, they modify their handed-

*The word *sinister* itself comes from the Latin for "left" or "on the left side," a term used in rites of augury for unlucky or unfavorable omens.

ness. Coren rejected this argument because it's just too difficult and unlikely for so many people to change their handedness, especially in older age. That left him with the extremely controversial elimination theory. Simply stated, lefties die off at an earlier age than righties.

It was a provocative idea—but was it true? Coren decided the best place to start looking for answers was somewhere that kept meticulous records of handedness and dates of birth and death. Sure enough, the search led him and psychologist Diane Halpern to *The Baseball Encyclopedia*. Examining 2,271 players, they found that lefties were "eliminated" sooner than righties. "For any given age," Coren wrote, "the percentage of left-handers who will die will run around 2 percent higher than the rate for right-handers." To emphasize the point, they noted that the oldest left-handed ballplayer lived to age 91 while the oldest right-hander made it to 109. "An 18-year difference!" Coren exclaimed.

Next, Coren and Halpern studied 2,875 California death certificates, questioned their families, and ended up with reliable information on 987 people. To their astonishment, they found that lefties die *nine* years earlier than righties. The gap was staggering. They also uncovered a significant gender difference. Left-handed women die four years and ten months sooner than right-handed women. But it's much worse for lefty men: Their life expectancy is ten years and one month shorter than for righties.

What could possibly explain these gaps? Coren and Halpern argued that lefties are prone to more serious health problems. They also argued lefties are 89 percent more likely to end up in accidents requiring medical attention and six times more likely to die in these situations. When their findings were published in the *New England*

Journal of Medicine in 1991, they were widely assailed. But follow-up studies—including an examination of top British cricket players—confirmed that lefties die sooner than righties, although the gaps were smaller. Today Coren concedes that his early research may have overestimated the deadliness of left-handedness, but he still contends that southpaws are eliminated sooner than righties. The consensus range, he says, is somewhere between two and five years.

If you're a lefty, fear not. For starters, you're a member of an illustrious group that includes Julius Caesar, Benjamin Franklin, Edward R. Murrow, Mark Twain, Charlie Chaplin, Oprah Winfrey, and five of the last seven US presidents. Perhaps more important, the world is becoming a safer place. Greater awareness of the accident risk for lefties has led to significant safety improvements. Industrial machinery and power tools have been modified so that righties *and* lefties can operate them. And for all the controversy over his findings, Stanley Coren is proud of his contribution to the world. "We've made the world a little bit safer for left-handers," he says. That's a good thing especially because his youngest son, one of five children, is a southpaw.

9

Hug the Monster

How Fear Can Save Your Life

On the prairies of southern Louisiana, there's a little town called Eunice, population 11,500. It's home to the Cajun Music Hall of Fame, a boisterous Mardi Gras festival, and the annual World Championship Crawfish Étouffée Cook-off. This is the corner of the South where Joe LeDoux grew up working in Boo's Meat Market, his dad's butcher shop. Boo was a rodeo rider, cowboy, and all-around tough guy who dispatched young Joe around town on a bike to sell pigs' feet for a nickel apiece. Boo let Joe keep the money. He also gave Joe another after-school job: plucking bullets from the wiggly brains of slaughtered cows. Stripping away the tough casing and membranes, Joe poked his fingers among the soft and slimy folds. "It just had a very nice feel to it," he says of the Jell-O sensation. He always wondered about the cows and what went through their minds—"to the extent that a cow has a mind"—as the bullet pierced their skulls. Back then, Joe was an altar boy who planned to join the priesthood. He never dreamed he would grow up to become a world-renowned neuroscientist and pioneer in the field of emotions and the brain.

Today the Empire State Building dominates the cityscape

from the window of LeDoux's corner office at New York University. It's a long way from Cajun country and cow brains. A computer monitor takes up most of his desk. The wide screen displays the title of a lecture he's writing: "Why Is It Hard for the Brain to Be Happy?" Joe—or Dr. Joseph LeDoux in academic circles—is an authority on how we create and retrieve memories of life's most significant and emotional events. If you squint, he looks like a skinnier version of David Crosby, the folk-rock legend. He's fifty-eight years old, and his thinning hair is brushed straight back. His sloping eyes are pale blue, his nose ends in a sharp point, and his graying goatee is neatly trimmed.

"We're just meat," the butcher's son says as we begin our conversation about the essential role of fear in survival. "Meat and bones and water," he goes on. "There's not a lot of mystery to that. That's what the physical body is. It's just a matter of how it's put together." The brain is "intricately wired meat," he continues. For years, most scientists believed that emotions were mental states that somehow occurred spontaneously in the brain. In this view, happiness or sadness just happen—*poof*—without conscious thought or influence by the brain's underlying structure. You get anxious or scared at a movie because fear materializes in your head independently of all the complex machinery in there. LeDoux is something of a revolutionary because he believes the opposite. The key to our emotions, he argues, is our brain circuitry. The way your brain is wired determines the way you feel. "It's the connectivity that makes everything happen," he says.

The emotional command center of the brain, LeDoux believes, is the amygdala, the nut-shaped clusters of neurons located a few inches from your ears. These little nuggets of gray matter, from the Greek word for almond, are your first and arguably most important survival tools. Every second of your life, they protect you from danger. They fire up your fight-or-flight response when you see a snake on a hiking trail or hear a noise in the night. More often, they trigger your emotional reactions to everyday things in life like the facial expressions of friend or foe. Your success as a survivor, LeDoux believes, depends on how you handle the daily torrent of emotions activated by your amygdala. As we've seen, fear drives some people to incredible action while it makes others freeze or freak out. Whether it proves to be one of your best survival tools or your downfall depends on your ability to control it. An effective survivor, LeDoux tells me, is someone who can shut off the fear alarm clanging in her head and channel the resulting motivation into purposeful action that reduces the danger. In short, "You have to be able to throw that switch from *I'm going to die* to *I'm going to survive.*" To see how this fear response circuitry works in a crisis, let's leave LeDoux's lab for a moment and join the revelry aboard the *Celebration,* a forty-eight-thousand-ton floating fun house owned by Carnival Cruise Lines.

1. The Man Who Fell Off the Cruise Ship

Tim Sears needed a vacation. Exhausted from his job as an industrial engineer, the thirty-one-year-old desperately wanted to escape the punishing winter in Lansing, Michigan. So Tim traveled to Galveston, Texas, with his best friend Mike for a weeklong booze cruise on the *Celebration.* After settling into their room, Tim and Mike wandered around the 733-foot ship, listened to a steel drummer, sipped a few drinks, ate a buffet dinner, and then split up in search of more fun. Tim made his way to the Galax-Z dance club, where he bought a few rounds for some new friends. Somewhere around midnight, he and Mike linked up, hung out briefly, and then separated again. That's all Tim says he remembers.

What happened next defies belief. Tim literally woke up in the ocean. Somehow, he had fallen overboard, plummeted around sixty feet, and blacked out on impact. When he regained consciousness, his first thought was simple: *What the hell is going on?* The ship was gone—there were no lights in sight—and he had no idea what had happened. "I felt completely sober," he says as he tried to make sense of "a pretty bad situation." He was wearing boxers, a T-shirt, and a sweatshirt. That's all. No shoes, pants, wallet, or identification, not that he needed any. Worst of all, he didn't have his glasses.

The fear alarm in Tim's brain should have been blaring, but he quickly shut it off. "I had just one of two options," he remembers. "I could either swim or sink, and I chose not to do the latter." His first concern was practical: He didn't want to get run over in the dark by a ship. So he set a goal: making it to the morning, when he would have a better chance of being spotted and rescued. "I never really had that sense of panic or fear," he says. "I never freaked out." Tim was a former paratrooper who had spent four years jumping out of planes with the Eighty-second Airborne Division at Fort Bragg. He was six foot one, two hundred pounds, and in pretty good shape. "I never really got that sense of being in a really screwed situation," he says. He knew he had to keep moving to stay warm. All night, he noticed twinkling lights in the distance and wondered if he was close to land. He learned later they were oil rigs about fifty miles offshore.

To stay afloat, Tim used every possible swimming stroke: back, breast, side, and dog paddle. He concentrated so hard on his mission that fear found no room in his brain. He didn't cry out to God, asking why this was happening. Instead, he accepted the new reality and dealt with it. *Well, I'm here,* he told himself. *If I'm going to get out of this situation, I need to keep maintaining and be strong and get through it at least until daylight when maybe I'll have a lot better chance to get out of here.*

If you were swimming for your life in the Gulf of Mexico in the middle of the night, you might panic about predators and other creatures of the deep. But not Tim. He kept his fear in check. He didn't even think about sharks or barracudas. When he encountered a school of jellyfish, the danger finally dawned on him. "That's when it kind of hit me that, oh yes, these aren't exactly the best waters to be swimming in." Like Brian Udell who ejected over the Atlantic, Tim even managed to take time to marvel at nature's beauty. Indeed, the most memorable experience of the entire ordeal, he says, came when "all of a sudden everything around me for as far as I could see were bright green little fish jumping." He stopped swimming just to look. "It was like, wow, this is amazing." At that precise moment, he thought, there was no place else in the world you could see such a stunning sight. "It was so beautiful."

Of course, Tim couldn't wait for the sun to rise. Riffing on the old banana boat song, he joked to himself: *Daylight come and me wanna go home.* After he'd been swimming all night, morning finally arrived, and he could see ships in the distance. His hopes picked up, but he was beginning to feel dehydrated. He swished seawater in his mouth for moisture and spit it out. A few times, he drank some ocean water, and promptly got sick. When none of the ships came near him after twelve hours, Tim felt the first twinges of desperation. He was badly sunburned, and his muscles ached. "I was getting real cold and really tired," he says. For the first time, a thought snuck into his mind: *I'm not going to make it.* Raised a Lutheran but not observant, Tim says he "made peace with God, and just said, 'It's okay. I've had a pretty good life. This is how it's going to end. Just take care of my friends and family because I don't know if they're going to find me.'"

Tim remembers closing his eyes and surrendering. He let seawater fill his lungs, and he began to sink. Floating downward, he felt completely at rest. Nothing hurt anymore. But around twenty feet below the waves, his eyes opened all of a sudden, and he started to swim toward the light. He can't explain why. He simply began to fight again. His voice cracks as he remembers the moment breaking the surface, choking and gasping for air. Coughing up water, he felt a new burst of energy. "I wasn't ready for it to be over with yet," he says. "I still had the desire and drive to live."

But the sun was getting low in the sky, and he knew there was no way he could survive another night. He had seen around twenty ships come and go and he prayed to God for any boat to save him. On the horizon, he saw a speck and watched it grow for twenty minutes. It was a cargo ship heading in his direction, and he swam toward it as fast as he could. He managed to get within 150 yards, but the waves were huge and he struggled to stay afloat. He stripped off the bright yellow MACKINAC ISLAND T-shirt from beneath his sweatshirt, tore it with his teeth to make a broader expanse of cloth, and waved and yelled as loudly as he could. Without his glasses, he couldn't tell if anyone had seen him, and the ship kept going. His spirits were crushed—but then the ship stopped. Tim saw a smaller boat heading back toward him. He shook his yellow shirt and screamed, "Help me! Save me!" Two sailors with a bright orange rescue pole drove right up to him.

Tim was rescued by a freighter called the *Eny* about fifty miles offshore. It was a Maltese-flagged ship that was ferrying copper from Chile to Texas. In its mess hall, they fed him plate after plate of food. No one spoke English except the captain, and Tim remembers the other sailors just stared at him like he was some kind of ghost. "And it just hit me," he says, "and I just broke down and started bawling, and one of the guys kind of patted me on the back like, *It's going to be okay.*"

As Tim gulped down food and water, the *Celebration* was already 450 miles away. No one on board the party boat even knew that he was missing. His buddy Mike assumed that he had hooked up with someone and was having fun behind closed doors. When the Coast Guard notified the Carnival ship that one of its passengers had fallen overboard seventeen hours earlier, the captain was stunned and incredulous. Tim wishes he could have seen his expression.

How does Tim explain his ability to survive for seventeen hours in the Gulf of Mexico? Was it purely a test of muscle and physical endurance? Not at all, he says. He believes he made it because of his ability to stay calm and focused. In Joseph LeDoux's lingo, he was able to shut off the alarm switch in his head and channel his fears into action. "I'm pretty sure if I had panicked and felt sorry for myself," he says, "I wouldn't be talking to you right now." Discipline and mental toughness were forged during his childhood in a strict religious school and also during his four years as a specialist wearing a maroon beret and baggy pants with the 82nd Airborne Division. He also believes in a certain amount of divine providence. Three months before the ordeal, he had started working out, losing twenty-five pounds. "That was just too much of a coincidence for me to not believe that this was all meant to be, that there's some kind of plan," he says. He hadn't gone to church in a while, but he also believes God must have heard his call.

Finally, he says, his instincts proved critical. For instance, he wore his sweatshirt the entire time in the water even though it would have been much easier to swim without it. "Looking back," he says, "it kept my entire upper body other than my head from getting burned. And that's stuff that I didn't even think about until afterward, but I just did it." Ingenuity also helped him through his trial. Swimming on his back, he navigated by following the jet streams in the sky. "I wasn't swimming in circles," he says. "I was going in a certain direction. You know little stuff like that actually helped me get through it because it gave me a goal, and I realized I was actually making progress and not spinning my wheels."

2. "It's All in a Nut"

Tucked away in Greenwich Village, New York, the Cornelia Street Café is a beloved spot where songwriter Suzanne Vega got her start, Monty Python performed skits in the 1980s, and Senator Eugene McCarthy recited poetry. Once a month, an unlikely crowd of self-described nerds and bona fide Nobel laureates gathers for a most unusual cabaret. John Nash, the schizophrenic Princeton mathematician who inspired *A Beautiful Mind,* stops by to listen to amateur acts like the Amygdaloids, perhaps the only rock blues band on earth made up entirely of neuroscientists. The group's name comes from those almond-shaped clusters in the brain.* "It's All in a Nut" is their theme song, an anthem written by Joe LeDoux, the NYU professor you met at the beginning of this chapter. In a gentle Louisiana drawl, he croons:

Why? Why do we feel so afraid?
Don't have to look very hard.
Don't get stuck in a rut.
Don't go looking too far.
It's all in a nut . . . in your brain.

LeDoux admits that his lyrics and vocals are no match for his heroes like Bob Dylan, Jimi Hendrix, and Eric Clapton. His performance, he concedes, is "mediocre at best," but getting better. Ever since he strummed his first guitar at age twelve,

* In October 2007, the Amygdaloids released a CD called *Heavy Mental.* The band has played at other venues including the John F. Kennedy Center in Washington, DC.

music has always helped him get closer to the truth. That's why he's proud of "It's All in a Nut," a ditty that encapsulates his belief that all of your fears, anxieties, and emotional memories originate right there in those little ganglia in your head.

LeDoux's lab at NYU's Center for Neural Science is a warren of rooms protected by keypads, card swipes, and a lot of security. There's a reason for all the precautions: It's no secret that they experiment on rats here. Indeed, much of what LeDoux knows about emotions and the brain comes from the unremarkable beige boxes arranged in neat rows in the lab. This is where rats are exposed to the twin tools of fear conditioning experiments: sound and shock. A rat hears a tone then experiences a short, mild jolt to its feet from a metal grille floor. It only takes one time before the rat begins to act afraid when it hears the sound. Waiting for the shock, it freezes with its fur standing up, its heart pounding, blood pressure soaring, and stress hormones pumping. LeDoux has carefully traced the different routes that the sound takes as it enters the rat's brain. He's also timed how fast it happens. He knows, for instance, that it takes about twelve one-thousandths of a second for the tone to reach the rat's amygdala through the so-called thalamic pathway. That's the quick and dirty route or what LeDoux calls the Low Road. It's fast, it's crude, and it tells the brain that there may be something dangerous ahead. But it's also imprecise. It triggers the fear alarm without really knowing the true nature of the challenge. This Low Road is twice as fast as the other route called the cortical pathway, which LeDoux has nicknamed the High Road. This circuit involves the brain's cortex, the furrowed gray matter where higher cognitive functions take place. The High Road is slower but it gives your amygdala more refined information to know if a threat is real.

So what does this all mean to the Survivors Club? Let's start with creepy crawlies. LeDoux is the Indiana Jones of neuroscientists. "I don't like snakes," he tells me. It's one of the reasons he's glad he left Cajun country, where slimy things slither everywhere. When he hikes with his wife in the Catskills, he proceeds cautiously, looking out for timber rattlers. If you walk along a trail and hear a rustling noise up ahead, LeDoux explains, the noise goes straight to your amygdala via the Low Road and your thalamus. You immediately freeze. Without thinking, your body protects itself. In survival terms, freezing is the optimum first response because it's not threatening to the snake and it buys you time to think and decide whether to throw a rock or run for your life. In the same instant that you freeze, the sound also travels up the High Road to your cortex where you can distinguish between the sound of a squirrel rummaging for nuts and a snake shaking its tail. As the Low Road works fast and furiously to save your life with an automatic reaction, the High Road sorts out the situation. The process is basically the same for sound and sight. When you step over a log and glimpse something long and skinny lying in the dirt, the information whips along the Low Road to your amygdala and you brace for danger. This automatic reaction—freezing—gives your High Road an extra split second to determine if it's a stick or a rattler. "If it is a snake," LeDoux writes in his brilliant book *The Emotional Brain,* "the amygdala is ahead of the game. From the point of view of survival, it is better to respond to potentially dangerous events as if they were in fact the real thing than to fail to respond. The cost of treating a stick as a snake is less, in the long run, than the cost of treating a snake as a stick." In other words, it's better to overreact to a piece of wood than to underreact to a venomous pit viper with retractable fangs.

LeDoux is a Low Road guy when he's hiking. So are most of us. Indeed, he describes e-mails from friends and fans who send snapshots of various sticks and branches. "I had a Low Road experience today," one of them writes, attaching a picture of a curved branch that LeDoux shows me on his computer screen. The key to survival, LeDoux argues, is controlling the

Low Road *reaction* with a more deliberate High Road *action*. In other words, LeDoux writes, "Those animals that can readily switch from automatic pilot to willful control have a tremendous advantage."

Some people, LeDoux believes, are innately more capable than others of throwing the switch from passive fear to active coping. But the rest of us, he contends, can learn to improve our responses to danger. How can he be so confident? Just watch his rats. Once the creatures are conditioned to fear a particular sound because of the shock that follows, LeDoux gives them the chance to move to a safe place when they hear the tone. If they make this choice, the shock doesn't happen. To protect itself from the jolt, the rat must take a small action. LeDoux has traced the precise way rats flip the switch from inaction to action. It turns out that each time they make an effort to protect themselves, they reinforce their coping skills. Each tiny step a rat takes away from fear makes it stronger and better prepared for the next time. LeDoux calls this "the neurological equivalent of getting on with life."

If rats can learn to control their fear and anxiety, he insists, so can the rest of us. Indeed, in the aftermath of September 11, LeDoux argued passionately that his rats actually give humans hope for being able to turn off the trauma and fear and resume our normal lives. LeDoux believes that the nonstop television imagery of the attacks on the World Trade Center reinforced our passive fear responses, rendering many of us "helpless and ultimately despondent." LeDoux argued that shutting off the TV, leaving the couch, and trying to resume normal life "removes you from those fear-arousing stimuli, and that's reinforcing and therefore allows you to take your next step."

These suggestions may sound like common sense, but LeDoux believes they're not as easy as they seem. Many people shut off the painful reminders of trauma by withdrawing from the world. To manage anxiety and fear, they close themselves off from everything and everyone. The challenge, LeDoux argues, is to tune out the painful reminders but to stay engaged in life. Our brains are "our best ally and our worst enemy," he writes. We're born with the equipment to make good decisions and learn from our mistakes, but we also worry and stress out more than any other animal. The best survivors either know innately how to manage their fears or they learn to exercise some control.*

Most of the scientists whom I interviewed for this book were surprised, even taken aback, when I asked them to apply their own theories to themselves. Is LeDoux good at throwing the switch? Is he able to manage his emotions? He hesitates at first but then tells a story about his own family trauma. In March 2005, his seventeen-year-old son Jacob died of an accidental heroin overdose in the family's downtown New York apartment. The teenager was a few months away from graduating from high school and going to New York University. He had struggled with substance abuse—he'd gone through rehab—but LeDoux and his wife were unaware that he was shooting heroin. The shock of Jacob's death was devastating, and LeDoux and his wife struggled for many months, wandering through life as if it were some kind of impenetrable fog. Somehow they carried on. LeDoux says making music helped him during the toughest times. His wife started to work on a project to help other at-risk youth. Some of their resilience is genetic, LeDoux believes. His wife comes from a "tough gene line" of Eastern European Jews, and his dad, Boo, was a hardy guy who lived to age ninety-two. It took time, but he and his wife were able to flip the switch

* Someday, LeDoux believes, we may be able to pop pills to erase traumatic memories. In a March 2007 study, he and his colleagues injected rats with a drug called U0126 and blocked the formation of specific traumatic memories related to beep tones that were followed by electric shocks. When the memories were stopped, the rats were no longer fearful of the sounds. A number of drugs are also being tested on people who experience horrible events like car crashes, assaults, and serious burns, and recent studies have shown the beta-blocker propranolol may help prevent terrible memories and thus reduce the traumatic stress response.

from despondency to coping and action. The key question is how quickly one can "extinguish" the negative emotional response. In his lab, it takes the rats only a few shocks to learn how to avoid unnecessary pain. LeDoux believes we can all teach ourselves to do the same.

3. What Does Fear Smell Like?

Dr. Denise Chen collects body odor. Not the stinky, sweaty kind. Chen is looking for something very specific under people's arms. She's hunting for the smell of fear, which may be critical in a survival situation. In pursuit of this quarry, she takes volunteers into her lab, stuffs gauze pads under their arms, and shows them thirteen minutes of a scary film. Her movie of choice: *Indiana Jones and the Temple of Doom.* She says people are always frightened by the bugs, snakes, and crocodiles. Then she takes the pads, puts them in glass jars, and freezes them at –112 degrees Fahrenheit. She repeats the exercise with a new group of people and shows them thirteen minutes of the comedy *Ace Ventura: Pet Detective.*

One week later, she removes the pads from the deep freeze and asks her subjects to sniff them. The challenge: Can they detect fear or happiness? An assistant professor of psychology at Rice University in Houston, Chen is a leader in the field of what's called the olfactory communication of emotion. It's the science of scent and feelings. Ants, bees, and worms communicate fear and alarm through so-called chemosignals or changes in body odor. Rats and mice, too. So what about humans? Do human body odors carry emotional information that can be detected by others? And can the smell of fear help you survive?

The answer to both questions seems to be yes, especially among women. When asked to choose between happy and fearful, a significant percentage were able to identify the smell of fright on the pads of men who watched the scary scenes from *Indiana Jones*. In addition, a substantial share were also able to detect the happy scent in men who watched the comedy. These findings were consistent with other research showing that women may be able to perceive emotional differences communicated through scent.

Chen was also curious about the impact of the fear scent on behavior. In the wild many fish, like trout, release chemical alarm signals when attacked or captured by predators, in order to warn other fish of danger. Do humans do the same, perhaps imperceptibly? If so, Chen wondered, what happens to our thinking abilities when we detect the smell of fear? Chen took her pads with fear sweat and put them on the upper lips of her subjects right below their nostrils. Then she gave them word association tests. When people smelled fear, their caution and vigilance increased. They behaved "as if they were motivated to avoid misses," Chen says. In terms of accuracy, the scent of fear improved test takers' performance by 5 percent over those who were given neutral pads, a meaningful difference in statistical terms.

In a crisis, Chen believes the smell of fear may help us interpret ambiguous situations when our other senses aren't giving us all the information we need. At an unconscious level, these smells may sensitize us to danger even if we're not aware of a threat.

To witness this mysterious process in action, let's turn to a warm summer evening in Amesbury, Massachusetts, a charming old mill town on the Powwow River. A thunder storm had passed through, and a full moon was shining. Adair Rowland had just attended a dinner party in her neighborhood and walked into the rambling Victorian home she had moved into with her young sons just eight weeks earlier. "As soon as I opened the door," she remembers, "there was something wrong." Instantly, she noticed the smell. "It was rank and I couldn't identify it," she says. She thought: *Maybe some rainwater from the storm had seeped into the house.* She bent down to pet her Siamese cat, who shrugged her off and darted away.

Upstairs in her bedroom, Adair kneeled on the rug next to an open window to check if it had gotten wet. That's when she noticed a yellow sweater knotted to the back leg of the bed. She untied the garment and stood up to turn on the lights, but the table lamps wouldn't work. Nervously, she moved from room to room. The day before, she had found another item of clothing fastened to a radiator pipe in the dining room. She had dismissed it as a game her kids were playing. But now she knew something wasn't right. Shaking with fear, she called her ex-husband and left a voicemail: "There's something wrong with the house. Call me or come here." Then she dialed her neighbor who said she would hurry over.

Still holding the cordless phone, Adair went to unlock the door when she heard a crashing sound on the back stairs. A bearded, barefoot man wearing only track shorts burst into the kitchen. He rushed at her with arms raised and what looked like bandages covering his hands. She immediately realized they were white socks and thought: *No fingerprints. He's going to kill me.*

Adair threw the phone at his face, but he grabbed her and started punching. She shouted that she had already called the police and her neighbor was on the way. He threw her against the cabinets and ran to lock the back door. Adair tried to escape through the sun porch, but he pulled her back into the kitchen. She managed to grab a frying pan and swing at his head, but he was too strong and fast. He knocked her to the floor and pressed his foot on her throat. Then someone started calling out at the door.

Crushing her neck, the attacker ordered Adair to make her friend leave. At first, she tried to offer him cash and the keys to her car. He answered: "Adair, I don't want your money. Now tell your friend to go away." When this stranger called her by first name, Adair feared the worst. She did exactly what he instructed, but her friend shouted that she was going to dial 911 anyway. Adair knew there wasn't time.

Now her attacker looped a long sash around her neck and jerked it tight. She recognized the scent and texture of the fabric from her junior prom dress in the attic. Lifting her up, her began to strangle her. Adair couldn't breathe and thought she would pass out. Then she saw a clear image of herself tied to her bed with tangled hair across her face. In this imaginary scene, she was dead. Suddenly, a voice cried out in her head: "No! This is stupid. My sons just lost their dad with the divorce. They can't have a dead mother. This is not me. It's the wrong script!"

Adair fought back furiously, and the intruder let go of the sash. She tumbled and hit her head on a radiator. When he reached out to help, she refused to touch him. As she stood up, the attacker calmly looked out the window and said, "They're already here." Adair could see the flashing blue lights of three police cruisers.

When her assailant tried to leave the house, he told the police he had been out for a jog, heard screams inside, and offered help. But the officers weren't fooled. Later, after his arrest, the white socks were found under the back seat of the police car, along with several condoms. Her neighbor identified the attacker as a resident of a nearby home. He was a married man with two kids who had recently been suspended from his job. At the trial one year later, he pleaded insanity but was convicted of attempted homicide and sent to prison for seventeen years.

Later that night and in the police investigation that followed, Adair learned more about the scent that she had detected upon opening her door. "It's the smell of fear," a police officer told her. From that first moment when she detected something rank, Adair's amygdala had fired into action. Her fear alarm went off, and she somehow managed to flip the switch, control her emotions, and call for help. When she saw the intruder, she immediately understood that her life was in danger and channeled her terror into fighting action. It's no exaggeration, Adair says, that fear and her maternal instinct saved her life.

But what if you can't shut off the alarms in your head? What if you're unable to transform fear into motivation? As you'll see in the next section, fear can keep you alive, but it can also kill you.

4. The Baskerville Effect

For a summer hike, Mount Doublehead in eastern New Hampshire is a beautiful spot with splendid twin peaks, rambling trails, and sparkling lakes. In August 2004, a thirteen-year-old named Antonio Hansell found himself on a six-mile walk in these woods. An inner-city kid from Boston, Antonio was spending a few weeks at a summer camp for underprivileged youth. By all accounts, he was a terrific boy who loved school, the Red Sox, and cars. From the very start that day, Antonio had fallen behind the other campers in his group. He and a counselor were making their way on the trail to catch up when they ran into a black bear. Terrified, they fled. In fact, Antonio took off so fast that he left a sneaker behind. Later, when he and the counselor returned for the shoe, the bear was still there. Once more they ran, and fortunately the animal didn't go after them. When they finally stopped running, Antonio told the counselor that he was too scared to keep hiking. Not long after, he started to have trouble breathing. Then he collapsed and died. An autopsy found nothing suspicious about his case. The medical examiner ruled that he had succumbed to natural causes.

The last time that a bear killed a person in the Granite State was back in 1784. This time, there was no mauling or even a drop of blood spilled. But the effect was the same. It appears that young Antonio was scared to death.* There's an elegant name for this phenomenon that comes from Sir Arthur Conan Doyle's mystery *The Hound of the Baskervilles,* in which a character suffers a fatal heart attack after being terrorized by a demonic dog. The Baskerville effect was discovered (and named) by Dr. David Phillips, a sociologist at the University of California, San Diego. Examining more than forty-seven million computerized death certificates, Phillips uncovered a very surprising pattern: On the fourth of every month, there's a spike in coronary-related fatalities among Americans of Japanese and Chinese ancestry. Surveying the United States, Phillips found 13 percent more Asian-American cardiac-related deaths on the fourth than expected. In California where these populations are concentrated, he discovered 27 percent more deaths.† In Mandarin, Cantonese, and Japanese, the words for "four" and "death" are almost identical, Dr. Phillips says, and many Asians are superstitious about the number. Indeed, in hospitals and hotels in the Far East, the number 4 is avoided just like the number 13 in parts of the Western world. Phillips tested and rejected all sorts of theories to account for the death peak on the fourth of every month. In the end, he concluded that fear connected to the number 4 was the only plausible explanation. "The Baskerville effect exists both in fact and in fiction," he declared in the *British Medical Journal.*

So how exactly does fear kill? And are you at risk? Dr. Martin Samuels of Harvard Medical School has collected quite a file on psychophysiological death, the technical term for dying of

---

* *Survival Tip:* If you run into a bear in the woods, experts say you should stay quiet, try to look as large as possible, avoid eye contact, and back away slowly. Your goal is to appease the bear and not threaten it. Never turn your back or show signs of fear or weakness. Whatever you do, never run away.

† In contrast with Chinese and Japanese Americans, Phillips says, Caucasians do not associate the number four with death and show no increase in cardiac mortality on the fourth of each month. Phillips also found no spike in "white mortality" on the thirteenth of every month, a date of alarm for superstitious people who suffer triskaidekaphobia or fear of the number 13.

fear or stress. As chief of neurology at Brigham and Women's Hospital in Boston, he keeps a slide show on his computer with newspaper stories about the members of his fright club. Why did a forty-nine-year-old woman and a four-year-old boy die suddenly in separate incidents on the Mission: Space thrill ride at Disney World? Why did a fifty-year-old Pittsburgh Steelers fanatic collapse from a heart attack seconds after his favorite player fumbled the ball trying to score late in the fourth quarter? Why did Kenneth Lay, the disgraced CEO of Enron, keel over from a heart attack while awaiting sentencing for his crimes? Why did Slobodan Milosevic, the butcher of the Balkans, die in his prison bed during his war crimes trial at the Hague? And why did young Antonio Hansell collapse after two brief encounters with a black bear in New Hampshire?

Despite such different circumstances, Dr. Samuels presumes, each of these victims was scared to death. "Under life-threatening stress without escape," he explains, "people will drop dead." Everyone is potentially at risk. "We all carry this little bomb inside us. If the situation is just right, if the stress is bad enough, if it's acute enough, if there's no way out, any of us can die." To understand how this works, Dr. Samuels says, you have to go back to the pioneering work of a legendary Harvard professor named Walter B. Cannon, who published his seminal findings on so-called voodoo death in 1942. Cannon studied cases around the world where people had supposedly died from black magic and curses. He outlined a scientific basis for these sudden, unexplained deaths, theorizing that the brain unleashed stress hormones that went directly to the heart, causing fatal arrhythmias. The release of these brain chemicals was the basis for what Cannon famously labeled the fight-or-flight response. Most of the time, it protects us in dangerous situations. In a minuscule fraction of cases, it kills. Over the years, Dr. Samuels has refined Cannon's theory by studying the actual hearts of dozens of victims of sudden unexpected death or SUD. He says their organs share many similarities, including massive injuries called lesions. If you're young and healthy, Dr. Samuels says, your heart is less susceptible to the deadly effects of fear. But it's also possible that you may already have a few minor nonlethal lesions on your heart from times you've been terrified. It's only natural.

The same physical process that can scare you to death has a fascinating flip side. It also turns out that you can be *thrilled* to death. Dr. Samuels points to a headline in the *Boston Herald* from November 12, 1994: "Golfer Dies After Perfect Hit." On the sixth hole of the Sun Valley Golf Course in Rehoboth, Massachusetts, a seventy-nine-year-old named Emil Kijek hit his first hole-in-one. On the very next hole, the retired dairy foreman grabbed his chest, said "Oh no," and collapsed. In rare instances, exhilaration kills, too.

* * *

Without a doubt, fear is the most ancient, efficient, and effective security system in the world. Over many thousands of years, our magnificently wired brains have sensed, reacted, and then acted upon every imaginable threat. Practically speaking, when you manage fear, your chances improve in almost every situation. But if your alarms go haywire, your odds plummet. For survival, then, here's the bottom line. If you're scared out of your mind, try to remember this air force mantra: *Hug the monster.* Wrap your arms around fear, wrestle it under control, and turn it into a driving force in your plan of attack. "Survival is not about bravery and heroics," award-winning journalist Laurence Gonzalez writes in his superb book *Deep Survival.* "Survivors aren't fearless. They *use* fear: They turn it into anger and focus." The good news is that you can learn to subdue the monster and extinguish some of the clanging bells. The more you practice, the easier it becomes. Indeed, with enough hugs, you can even tame the beast and turn him into your best fried and most dependable ally.

SURVIVAL SECRETS

Who Gets Lost (and Found)?

In July 1986, a nine-year-old boy named Andy Warburton told his mother that he was going for a swim in nearby Tucker Lake. Andy and his family were vacationing in the town of Beaver Bank, nestled in the thick forests of Nova Scotia not far from Halifax. In the world of search and rescue, this rugged and remote area on Canada's south-eastern coast was once known as the Lost Person Capital of the World. People went missing here all the time, gob-bled up by the woods.

Andy never made it to the lake, a short distance from his aunt and uncle's home, and for the next eight days searchers swarmed the wilderness, hunting for a blond boy in a blue-and-white shirt and green shorts. Small chil-dren are often afraid of searchers and fearful they'll be punished for getting lost, so Andy's mother was brought into the woods and given a megaphone. "Mommy and Daddy are waiting for you," she cried out. "The men will not hurt you. They are here to help you. Please come out on the road."

Expert trackers say that an average person leaves behind two thousand clues every mile he travels. Each step you take produces evidence—a footprint, a broken twig, a clump of mud, a twisted blade of grass. A team of well-trained searchers spaced ten feet apart usually can detect 95 percent of the useful clues. If they're spaced fifty feet apart, they'll discover 75 percent of the clues. And yet Andy War-burton disappeared. Searchers turned up nothing useful. As the days went by, authorities even flew psychics over the forests hoping for supernatural help. It was the largest search in North American history, with more than fifty-eight hundred volunteers plowing through dense brush.

During this massive operation, a young professor named Ken Hill received a phone call from the Royal Canadian Mounted Police asking for help. The Mounties wanted to understand how a lost boy might behave in the woods. Hill was the only child psychologist at St. Mary's University in Halifax, and as he tells me this story, he's sitting in the same office where that one phone call changed his life. He had followed the search for Andy on the local news and was glad to drive over to the command post. When he arrived in Beaver Bank, he was overwhelmed by the crowds of volunteers and the chaos of the scene. The Mounties took him to the search map, where they identified Andy's Point Last Seen or PLS. Hill remembers staring blankly at the topographic map. He could barely understand its markings and had no ideas to offer the Mounties. He knew nothing about lost boys and their behavior. "It felt like I let Andy down," he recalls. So he went outside and joined up with a search party to scour the woods.

After 165½ hours of search operations, two men found Andy's body in Rasley Meadow, curled up in the fetal position with his legs badly scratched up. They radioed a special code to the command post. The boy was dead. Alone and terrified, he had obviously run helter-skelter through the woods. His body was less than two miles from his aunt and uncle's home. He had wandered farther than search organizers predicted. They had never even thought of looking in the meadow.

Ken Hill was haunted by his failure. At age forty-one, he took it personally. He had a daughter the same age as

Andy, and as a boy he had gotten lost for a day in the Santa Monica Mountains of Southern California. So one week after Andy Warburton was buried, Hill applied to join the local search-and-rescue team. He also scoured the literature on so-called lost person behavior and was stunned to discover the work of a man named William Syrotuck, a pioneer in the field who had compiled data on lost people in the United States. Syrotuck seemed to have all the answers, including the probability zones where different kinds of lost persons would be found. Indeed, Andy's body was discovered close to the average distance that Syrotuck predicted for nine-year-old children.

Hill learned that all you need to know about lost people is their age and their outdoor activity and you can calculate how far they're likely to travel and where you should go looking. You don't need a detailed personal history or psychological profile. You just need some basic information. "It is more important to realize that a known percentage of all lost persons is found within a one- or two-mile radius than it is to know how they got there," he writes. Hill has interviewed countless people who have gone missing in the woods. He's looked into their wide eyes and listened to their trembling voices. "Fear is the enemy," he says. "Fear activates the large muscles in the legs." Lost people want to run. They also lose their heads and sometimes forget even to look in their own backpacks for food and water.

It turns out that no matter where you are in the world, lost people behave in the same way. Who you are determines how far you're likely to wander. In Hill's research, hunters get lost the most often. Typically, when they're found, they've traveled between 0.94 and 2.25 miles. Hunters are usually in pretty good physical condition, but when they're lost, they often push beyond their abilities. Yet they're also easy to find in a search. They're good communicators and they typically have solid outdoor skills. Hikers are another big group that gets lost. They're very dependent on trails and most often don't have maps or compasses. When they're found, they've typically traveled between 0.87 and 2.88 miles. Small children between ages one and six usually travel between 0.67 and 1.65 miles. The smallest ones, between one and three, have no idea they're lost. If they're separated from their parents, they have no ability to find their way; they wander aimlessly, and they typically don't go very far. They're usually found sleeping. Naturally, three- to six-year-olds are more mobile, and they understand the idea of getting lost. They tend to take care of themselves better than older children and even adults. They burrow in bad weather by sleeping in caves or hollows. Typically, they're "stranger resistant," meaning they won't respond to searchers. Older children between ages seven and twelve will run when they're lost. The distance is usually between 0.92 and 1.70 miles. They're often afraid of punishment, and they won't answer searchers until they're cold and hungry. They have the same fears as adults, only more acute. Of all the different groups, it turns out older people are excellent survivors because they tend to build shelters and await rescue.

So what should you do if you get lost on a hike? Hill's number one survival tip is to stay where you are or find an open place nearby. Of the eight hundred Nova Scotia lost person reports that he reviewed, only two intentionally stayed in one place in order for searchers to find them more easily. "We always find clues before we find the victim," he says. "What does that tell you?" Hill's point is that if the victims had stayed in one place, they would probably have been found sooner.

Incredibly, whenever people get lost anywhere in the world, search-and-rescue officials turn to the same probability circles. And thanks to Ken Hill's work, Nova Scotia is no longer the Lost Person Capital of the World.* Children throughout the province are taught "woodsproofing," and hunting licenses now require a working knowledge of navigation with a compass. Still, when Hill teaches classes on lost person behavior, he lays out the facts of the case that still haunts him and he challenges his students: "Try to find Andy Warburton." Using the probability zones, it's easy to narrow down where to look. Two decades ago, locating a lost person was something of an art. Now it's a science. I ask Hill how long it would take to find Andy Warburton if he went missing today. "First day," he replies, "first few hours, no question."

---

*The dubious distinction now belongs to British Columbia, according to Hill.

10

Too Mean to Die

Does the Will to Live Make Any Difference?

How the hell did I survive that night?

Nineteen years later, Trisha Meili still wonders why she's alive. "I don't think I can explain it," she says, "but it's okay that I don't understand."

It was Wednesday, April 19, 1989. Two policemen were sitting in an unmarked car on the road that cuts across 102nd Street near the top of New York's Central Park. Suddenly a couple of men rushed up to say they had seen a guy in the woods who was badly beaten and tied up. The cops immediately went looking and discovered a body, but it wasn't a man's. Covered in mud and blood, it was a naked woman, lying in a ditch not far from the road. Her running tights had been ripped off and her bra was shoved above her breasts. Her shirt had been used to gag her mouth and tie her hands together in front of her face like she was praying. Her skin was purple and black, and she was in shock. Bleeding badly from five deep slashes on her head, her body was shaking and jerking, and she was unable to respond to the officers' questions. An ambulance rushed her to Metropolitan Hospital, where her core body temperature measured eighty-five degrees. She had lost 75 to 80 percent of her blood, barely had a pulse, and her blood pressure was undetectable. She was experiencing Class 4 hemorrhagic shock, the most severe kind. Her skull was fractured and she was bruised on every inch of her hundred-pound body except for the soles of her feet. Her prognosis was so grim that a priest was summoned to give last rites. On the gurney, she flailed so violently that nurses tied her down with restraints. Some people wanted to believe her thrashing was proof that she was fighting for her life. In fact, it was a sign of serious brain damage. The Central Park Jogger—as she would come to be known—was twenty-eight years old with a promising career at Salomon Brothers, one of Wall Street's elite investment banks. Raped, beaten, and left for dead, she was barely clinging to life. One doctor even told her family, "It might be better for all if Trisha died."

On the coldest, windiest day of the year not far from New York's Grand Central Station, Trisha steps through the revolving door. Wearing a dark blue suit with velvet lapels, she's pretty and petite with hazel eyes and only the slightest scar on her left cheek that's deftly concealed in a dimple crafted by a surgeon. The Central Park Jogger is forty-eight now, and her grace and sparkle remind me of Julie Andrews. Trisha doesn't remember the attack in Central Park. In fact, she has no memories of what she calls her "six-week plunge into darkness." She can only recall a dinner invitation from a co-worker that night. The rest is all blank—the brutal attack, twelve days unconscious in the ICU, six more weeks "delirious" in the hospital, a city and nation gripped by her trauma, and an outpouring of sympathy that included eighteen pink roses from Frank Sinatra.

"I can only believe that I was scared out of my mind," she says of that night in the park. "I was told by the police that I fought." She raises her hands as if to show the bruises they found on her hands from punching her assailants. "I was a fighter," she says. Trisha marvels that she was left to die all alone in a wet and muddy ditch. "I was just lying there" for three hours or more, she says, "and no one knew I was there." Given her injuries, Trisha knows she should have perished. "It makes me trust that there's something else going on," she says. Over the years, she has searched for answers beyond science. Was it God's grace or an unbreakable will to live? "I don't know," she says. "It's hard to pinpoint one single reason." Long after the attack, Trisha went to see Dr. Robert Kurtz, who directed the team that saved her life at Metropolitan Hospital. She wanted to know his expert opinion on how she survived. "You're an indomitable person," Dr. Kurtz told her. "You were in a situation where other people like you might well be dead—and you weren't." Trisha ruminated about Dr. Kurtz's use of the word *indomitable*. The idea "can't be defined in medical terms," she later wrote in her best-selling memoir, *I Am the Central Park Jogger: A Story of Hope and Possibility*. "The phrase sounds strange coming from a scientist, and perhaps that's why it resonates strongly in me. Others have called me 'indomitable,' but it's not an image I've ever had of myself. Now I think, *Maybe*. Even in the coma I was apparently goddamned if I was going to die."

Goddamned if I was going to die. That's probably the most succinct definition of the will to live that I encountered in all of my interviews with survivors and experts. *The will to survive* is a ubiquitous phrase that pops up whenever there's a survival story in the news. If you can name a disaster, chances are that reporters found a living and breathing example of the will to survive. Not surprisingly, almost every survival guide ranks the will to live at the top of the list of critical tools. In October 1957, for instance, the US Army produced a little beige booklet called FM 21-76. It's the *Field Manual for Survival*, the official instructions on how to stay alive in hostile environments. In its very first pages, the book tells soldiers that the experience of hundreds of servicemen in World War II and Korea "prove that survival is largely a matter of mental outlook, with the will to survive the deciding factor." If you lose your will, you're finished. "Without a 'will to survive,' your chances of surviving are greatly diminished," adds the FAA's *Basic Survival Skills for Aviation*.

But does the will to live really make a difference? I turned to survivors and science for answers, and what I found may surprise you. As you'll see in the pages ahead, the will to live is a mysterious, ineffable force that defies rigorous definition and measurement. It may originate with an electrical impulse in the brain or it may be delivered by the hand of God. Whatever it is and wherever it comes from, one thing is certain: Your attitude can be your life preserver, so hold on tight.

1. The Fighting Spirit

In October 1979, a British psychiatrist named Dr. Steven Greer rocked the medical world with a study in *The Lancet,* a leading medical journal. Dr. Greer found that women with breast cancer who demonstrated a "fighting spirit" survived longer. The study took place at King's College Hospital in London, and Dr. Greer grouped sixty-nine women into four categories: Fighting Spirit; Denial; Helpless and Hopeless; and Stoic Acceptance. After five years of study, 75 percent of the Fighting Spirit and Denial groups had "a favorable outcome" compared with just 35 percent of the Helpless and Stoic groups. The 40 percent difference seemed stunning because no one had ever proven that the mind could triumph over the body or that psychology could prevail over biology. Of the women in this study who ultimately died, 88 percent fell into the Helpless and Stoic groups while only 46 percent of the women who stayed alive and cancer-free demonstrated these attitudes.

Twenty years later, Dr. Greer co-authored a larger study in

The Lancet of 578 women, but this time the results were far less encouraging. Researchers "could find no survival difference when comparing one type of psychological response with another." Specifically, they were unable to uncover any evidence that a fighting spirit makes you live longer. "Our findings suggest that women can be relieved of the burden of guilt that occurs when they find it difficult to maintain a fighting spirit," they wrote. On the other hand, they argued, persistent feelings of helplessness and hopelessness—the opposite of the fighting spirit—are bad for your health. Of those women who expressed beliefs like "I feel like giving up," only 58 percent were alive or in remission after five years, compared with 72 percent who did not feel hopeless or helpless. The authors concluded that the fighting spirit won't prolong your life, but helplessness and hopelessness may actually shorten it. "It is not what may be *added in* by fighting but what is *taken away* by being helpless that seems important in disease outcome," they wrote.

Dr. Greer is eighty years old now, semi-retired, and works one day a week as a psychiatrist at St. Raphael's Hospice in a village called Cheam in southwest London. He treats terminally ill patients experiencing anxiety and depression. He may be one of the pioneering authorities on the fighting spirit, but he doesn't sound like a man who's very convinced anymore about the power of the mind. He tells me that the fighting spirit is very hard to measure—that's why studies fail to prove its impact—and that a positive mental attitude exerts only "a modest effect" on the outcome of patients with cancer. Indeed, he says, the fighting spirit is "much weaker than biology." While the fighting spirit may not influence your *quantity* or length of life, it will certainly improve the *quality* of your remaining time. People with the fighting spirit show "far far less" depression, anxiety, and other mental health problems, he says. I ask whether he has the fighting spirit in his own life. The question amuses him. "Yes, I hope so," he says. Then he pauses. "I haven't had cancer. So I can't be sure."

* * *

When Samuel Gruber was twenty years old, he paid seven dollars for a ticket on the *Blue Goose,* a World War II–era sailboat that crisscrossed the waters off Miami Beach. He brought along his mask, snorkel, and spear gun for some fishing in the reefs around Fowey Rocks. As he took aim at a grouper under a rock, he remembers the most terrifying sight. "I saw this submarine," he says with a laugh. In fact, it was a huge hammerhead shark. It may have looked five hundred feet long, but Gruber says it was probably more like fourteen and now it was only five yards away. "I thought I was dead," he says.

Today Gruber is perhaps the world's greatest authority on sharks. He's lived with the predators for most of the past fifty years, focusing on one species called the lemon shark, known for its yellowish belly. On the tiny island of South Bimini in the Bahamas, you'll find "Doc" at his field research station, a rickety old shack known as the Shark Lab. Doc is an emeritus professor at the University of Miami, where he started teaching marine biology in 1960. A formidable man with an aggressive intellect and fierce devotion to his sharks, he has the leathery look of a veteran surfer with a woolly beard and graying hair swept straight back.

If you want to understand survival, Doc explains, sharks are a magnificent case study. As a type, they're exquisitely engineered to live forever, he believes, unless humankind kills them all off.* In one shape or another, they've been around for 450 million years—that's well before dinosaurs roamed the earth. Why have sharks survived so long? It's simple, he says. They've adapted incredibly well to change and they're very good at learning. In other words, "they're built to work really well."

* One hundred million sharks are slaughtered every year around the world. In the past two decades, overfishing has resulted in a 90 percent decline in many shark populations and a 98 percent drop in species like the scalloped hammerhead along the East Coast of the United States.

They've got around three hundred teeth arranged in five rows for feeding and fighting, and they're all replaceable. A typical shark can go through fifty thousand teeth in a lifetime. Sharks are also remarkably energy-efficient, with a skin coating that allows them to glide 20 percent more than other fish. They're equipped with extra senses that humans and other sea creatures don't have so they can detect electrical fields to navigate and hunt their prey. They've got excellent smell and vision and extraordinary immune systems and genes that are resistant to mutations like cancer. They even come equipped with a built-in antibiotic called squalamine that helps them resist infections. And above all, they have big brains. "I learned to speak to sharks," Doc tells me, explaining that he trained them to answer very simple questions by winking. Indeed, they're eighty times faster than rabbits and even cats at learning simple visual tasks. They also seem able to remember what they learn for a year or more. "Strength is not necessarily the key to survival," he says. "You also need to be smart." Doc pinches the wrinkled bronze skin on his wrist to show how thin it is. "We're not well designed," he says about the human species. "It's hard to believe that we can survive at all." After all, "we don't have claws." So we made up for our design deficiencies with our brains. Indeed, we conquered this planet—and have survived this long—because of our ability to think, learn, and remember. To a certain degree, Doc says, we humans are just like sharks without the teeth.

To be sure, Doc himself is very much like the magnificent survivors that he studies. And that's the real reason I've come to see him: He's got the fighting spirit. Even his friends say he's too mean to die. At the age of thirty-seven, married with two young children, he was diagnosed with non-Hodgkin's B-cell lymphoma. Determined to see his kids graduate from high school, he did everything his physicians told him. Old-fashioned chemotherapy sent his disease into remission, and he followed an incredibly healthy lifestyle, raising goats for milk and bees for honey. He meditated using so-called guided imagery, envisioning his white blood cells as great white sharks swimming through his body devouring cancer cells. But in 1982, the lymphoma came "roaring back." The years that followed were miserable. All kinds of treatments failed to stop the disease, and by 1989, he had run out of options. His disease seemed unstoppable, and after he sought advice at the Dana-Farber Cancer Institute in Boston, the experts sent him home to Florida to prepare for death. There was nothing they could do. They recommended he write his will and settle his debts. His own doctor in Miami also refused to give him chemotherapy. It was too late. There was no point trying anymore.

But Doc refused to give up. With the help of a journalist in Florida, he used the early Internet to search for something—anything—to save his life. In the footnote of a scientific paper, he found a glimmer of hope. Eleven late-stage lymphoma patients had been given a chemotherapy drug called fludarabine, and three had responded well. Sure, it was a tiny number, but Doc was ready to try anything. He told his physician: "I want you to give me fludarabine." The doctor replied, "It's not for you." So Doc showed him the footnote in the science paper, and the oncologist was surprised. "I didn't know that!" he said. In 1990, Doc received his first intravenous treatments of fludarabine, and soon after the lymphoma disappeared. "That was the last problem I ever had," he says. "The rest is history."

In Doc's case, the will to live meant the determination to find his own cure even after the experts had sent him home to die. That life force didn't just appear when he was diagnosed with cancer. It's reflected every day at the Shark Lab in Bimini. His friends and students say he's tough, single-minded, and ruthlessly focused on what needs to get done. His colleagues joke that even a shark wouldn't go near him for lunch. Doc says his triumph over lymphoma is explainable in two parts: the fighting spirit *and* biology. When the cancer experts gave up on him, he didn't curl up and surrender. Instead, he went to work. "You've gotta watch out for doctors," he says. "They can do you in." Doc believes that all patients need to become experts on their own illnesses. He has no doubt that by learning everything possible about your challenge, you can make a difference and even save your life. Today he marvels that fludarabine is a well-known and accepted treatment for lymphoma. The second key to his success involves biology, a factor beyond his control. Why did his body respond to the experimental treatment? Doc says it defies explanation. It turns out his body has one hundred times the normal level of natural killer or NK cells that can help destroy tumor cells and others infected with viruses. They're a critical part of our immune systems, and Doc doesn't know why he's got so many. He jokes that maybe it comes from spending so much of his life cavorting with sharks. Then he scratches his bristly beard, shakes his head, and grins.

2. Can You Kill the Will to Live?

On May 22, 1992, Mary Ward awoke in the middle of the night when she heard someone splashing in the pool outside her bedroom window in Stanton, California, a small city on the road to Disneyland and the beaches of Orange County. The seventy-nine-year-old widow lived alone in the apartment complex where she also worked as an assistant manager. When she went outside to investigate the ruckus, she found Jose Alonso Garcia in the water. He was twenty years old, a dishwasher at a local restaurant, and very drunk. Mary ordered him to leave, but when he made an aggressive move, she fled back to her apartment. While she was dialing 911, Garcia entered her home and attacked violently. The desperate call was captured on tape: "Please! Help! He's trying to choke me . . . Hurry!" Mary cried. "He's trying to rape me! Please! Help!" When sheriff's deputies arrived, they found Garcia still on top of Mary and forcibly pulled him off. His blood alcohol content was 0.22—almost three times the legal limit.

Mary survived the attack—her physical injuries were treated immediately—but when Orange County deputy district attorney David LaBahn met with her to prepare for the rape trial, he found a woman whose spirit was broken. "I knew it was a very, very grave situation," LaBahn remembers. There were ligature marks on her neck and a scab on her nose. She was not really there. She didn't seem to want to live anymore. Sure enough, Mary died the following month. When he heard the news, LaBahn immediately thought: *He killed her. If she hadn't been attacked, she wouldn't have died.*

So LaBahn made a bold decision: He would prosecute Garcia for homicide as well as rape. He would argue that the illegal immigrant murdered his victim by destroying her will to live. Before the attack, she was an energetic, feisty, fun-loving Southerner who enjoyed entertaining her friends. After the assault, she was an isolated and weeping wreck. When LaBahn brought his theory and case in front of the grand jury, they literally applauded. The decision was unanimous—nineteen to zero—to charge Garcia with murder.

During the ensuing trial, LaBahn argued passionately that when Garcia killed Mary's will to live, it was tantamount to murder. "These terribly atrocious acts led to [Mary's] death," he told the twelve jurors. "There is a direct causal link between [Garcia's] actions and her death." LaBahn summoned a variety of expert witnesses who testified about the science of willpower. At the conclusion of the trial, he argued that even if Garcia's heinous attack shortened Mary's life by only one second, it was still homicide.

The courtroom drama in Southern California drew international media attention because it was the first legal case to explore whether your state of mind can really influence the timing of your death. The defense team ridiculed the murder charge as "voodoo." They conceded that Garcia had raped Mary after drinking twelve beers and a bottle of wine, but they insisted he was not a killer. The cause of her death, they argued, was pneumonia brought on by advanced lung cancer and the failure of her heart and kidneys. After almost four days of heated deliberations, the jury of seven men and five women agreed with the defense. They convicted Garcia of rape but not homicide. Mary's death, they believed, was the result of her advanced illnesses, not the breaking of her spirit. Garcia was found guilty of five felony counts of sexual assault and was sentenced to twenty-eight years in state prison.

LaBahn is now an executive with the National District Attorneys Association in Alexandria, Virginia. To this day, he insists "the will to live is real," and the novel legal theory was simply "before its time." So what exactly is the will to live—and if it exists, where might it originate? To find answers, you must start by looking inside your head.

Your brain is a three-pound generator that runs on glucose, or sugar, transported around in your bloodstream. If you could tap all of the electricity released by the hundred billion neurons in your brain, you could power a ten- or fifteen-watt flashlight. That may not sound like much, but even when you're asleep, your noggin pulses with brain waves. To pinpoint the origins of the will to live, you need to track these electrical sparks or impulses, says Dr. Ken Kamler, the adventurer from chapter 5 who specializes in extreme medicine. "Thoughts contain electricity," he writes in *Surviving the Extremes*. "Strong thoughts contain more electricity." A person fighting for his life generates quite an electrical storm in the brain, Dr. Kamler says. In his view, the will to live originates in the cortex—or gray matter—of the brain. Using advanced brain scans, he says, scientists can actually trace a person's thoughts by following blood flow all the way to the place where decisions are made. They can then track the blood as it changes directions and heads back into parts of the brain where choices turn into action. Dr. Kamler says the brain's "commander in chief" is called the anterior cingulate gyrus or cingulate. It's the decision-making center where you choose whether to live or die, fight or surrender, go forward or retreat. If your cingulate works well, you make decisions and act on them. If your cingulate is damaged, he notes, you're likely to languish "in a state of placid indifference."

Neurologists agree with Dr. Kamler that the cingulate cortex acts as the interface between the cognitive and emotional centers of the brain. But they don't all concur that the will to live emanates from one particular spot in your head. Decisions are made in so many different places in the brain, they say, and each person's circuitry works differently. But debating the precise location of willpower in the brain misses the larger point that Dr. Kamler says he's trying to make. The real question is: What causes these electrical impulses and decisions in the first place? "Is it an electrobiochemical resource hidden deep within the brain," Dr. Kamler asks, "or a force instilled from without by a higher power?" The electrical impulses streaming from the cingulate "may be the origin of will," he argues, or they may be "the first detectable result of faith." In the end, it's a mystery, he says. "The answer is within our bodies but beyond our grasp." For all the brain scans in the world, Dr. Kamler believes that "the fundamental nature of the human will must remain unknowable."

During those grim days when the Central Park Jogger was in a deep coma, doctors told her family and friends that when Trisha ultimately woke up, the most dominant aspect of her personality would emerge first. Apparently, that's the pattern when coma patients regain consciousness: Their strongest personality traits are most apparent. So Trisha's family discussed which qualities they expected to see first. Their opinion was unanimous. "We all agreed it would be your determination," they told her later.

I ask Trisha if determination was indeed her most important survival tool. She pauses for a moment, looking for another word. "Is it persistence?" she asks. Her eyes search the room. She hasn't nailed the thought, and she seems to be trying again. "My husband would say I'm stubborn, too. And I am!" Trisha adds that a lot of other factors contributed to her survival and recovery. Her parents, aunt, two brothers, and friends cared about her deeply. She was a Phi Beta Kappa graduate from Wellesley College and earned two master's degrees from Yale. She was focused, ambitious, and competitive. She was also kind, loving, and sensitive. She received incredible support and prayers from friends and strangers around the world. All of these factors, Trisha writes, "allowed me to survive the attack and kept my mind concentrating on my recovery rather than indulging in self-pity or becoming obsessed with destructive rage." She continues: "As fire turns sand to glass, so the attack forged a new me. The basic elements remained the same, but in a different form."

It may sound strange, but Trisha also believes that her massive brain injuries may actually have made her recovery easier. As we discovered in chapter 4 on life and death in the ER, young and healthy people have a much better chance of recuperating from brain trauma. In Trisha's case, the damage was so great that it actually ended up helping her heal. For starters, she was unable to remember the actual attack and was never shattered by flashbacks or nightmares. It was an unexpected blessing: All of those horrors in Central Park were simply erased from her mind. The trauma also limited Trisha's concept of the future because she didn't have the capacity or inclination to worry about the road ahead. It took all of her energy and concentration to deal with the hour-to-hour work of healing and living.

There was simply no time to obsess about whether she would fully recover. Salomon Brothers promised that she could have a job when she was ready, so there was no need to panic about her career. In short, there was no past or future tense for Trisha. It was all present. With such serious physical and cognitive limitations, she focused everything on the here and now. The damage, she came to believe, was a kind of gift. "Paying attention to what was right in front of me," she says, "that was all my brain could do." Today, in many ways, she feels stronger than ever. Some things take her a bit longer—and her balance isn't perfect—but she's grateful for her health and she treats her body as well as she can. She exercises but not compulsively, running five to ten miles per week. She also lifts weights, practices yoga, cycles, and kayaks. She's married, lives in Connecticut, and travels around the country giving motivational speeches. The focus of her talks is hope, possibility, resilience, and what she calls "the power of the present moment."

3. Dragons and Magic Charms

Alice Stewart Trillin was a beautiful, brilliant, and optimistic woman who seemed to float around New York City, teaching college courses in writing, producing award-winning public television programs, raising two young daughters, and serving as muse, foil, and in-house editor for her husband, Calvin Trillin, the writer. In 1976 at the age of thirty-eight, Alice was diagnosed with lung cancer. She had never smoked and believed the disease was probably brought on by exposure to her parents' cigarettes. Her doctor told her she had a 10 percent chance of survival.

In March 1981, after enduring surgery, radiation, and chemotherapy, Trillin wrote a poignant essay in the *New England Journal of Medicine* called "Of Dragons and Garden Peas: A

Cancer Patient Talks to Doctors." It was an impassioned argument about what it's like to live in a place that she called "The Land of the Sick People" where she and so many others faced "the dragon"—cancer—every day. Trillin's essay was a powerful exploration of the will to live, which she considered a talisman or magic charm that's supposed to ward off danger and evil. For people with cancer, she wrote, the talisman of will was supposed to keep the dragon away. The more you wanted to live, the more you could beat cancer. Trillin described how she and many other cancer patients embraced every story about people whose will to live forced their illnesses into retreat.

Of course, Trillin wrote, there are perils to believing in this kind of magic. She described one friend who "wanted to live more than anyone I have ever known" but who succumbed to cancer. "The talisman of will didn't work for her," she wrote. "I know that believing in [will] too much can lead to another kind of deception . . . If I get sick, does that mean that my will to live isn't strong enough? Is being sick a moral and psychological failure? If I feel successful, as if I had slain a dragon, because I am well, should I feel guilty, as if I have failed, if I get sick?"

Trillin was a woman with an extraordinary, exemplary will to live. "The strength of my love for my children, my husband, my life, even my garden peas has probably been more important than anything else in keeping me alive," she wrote. "The intensity of this love is also what makes me so terrified of dying." Trillin was also practical and realistic. She knew the dragon was always lurking, but somehow, that menace made her life even richer. In her essay, she quoted the cultural anthropologist Ernest Becker: "I think that taking life seriously means something such as this: that whatever man does on this planet has to be done in the lived truth of the terror of creation, of the grotesque, of the rumble of panic underneath everything. Otherwise it is false."

"We will never kill the dragon," Trillin concluded. "But each morning we confront him. Then we give our children breakfast, perhaps put a bit more mulch on the peas, and hope that we can convince the dragon to stay away for a while longer." Trillin survived for twenty-five years after her diagnosis. On September 11, 2001, the same day that al-Qaeda attacked America, she died at New York Presbyterian Hospital. In the end, it wasn't the dragon that took her life. The cause of death was congestive heart failure brought on by damage from her radiation treatments. The cure for her cancer had killed her. She was sixty-three years old.

4. Can You Postpone Your Death?

Westminster Abbey in London is an unlikely place to go searching for the will to live. Since 1066, British monarchs have been crowned and buried in the great Gothic church near the Houses of Parliament and the River Thames. The list of those entombed in the abbey is long and illustrious. King Henry III, Queen Elizabeth I, and Queen Mary I are here. So, too, are Geoffrey Chaucer, Charles Darwin, Charles Dickens, Rudyard Kipling, Dr. Samuel Johnson, and Sir Laurence Olivier. The abbey is so crowded with three thousand souls that the playwright and satirist Ben Jonson was buried upright in a standing position in Poets' Corner.

As a young sociologist, Dr. David Phillips wandered the hallowed nave, transepts, and cloisters of the abbey, making notes on the birthdays and death days of the luminaries. You met Phillips in the last chapter—he discovered the Baskerville effect. For years, Phillips has been fascinated by the will to live and whether people can postpone their deaths in order to participate in important social events like birthdays, holidays, and even political elections. When he analyzed the data, Phillips found something quite surprising. Famous people seemed able to delay their deaths until after the anniversary of their birth. More specifically, he observed a noticeable dip in deaths before famous birthdays and a peak afterward. Expanding his research beyond London, Phillips looked at 1,333 of the best-known people in history, including George Washington, Benjamin Franklin, Thomas Jefferson, Mark Twain, and Thomas Edison. "The more famous the group," he wrote in the *American Sociological Review,* "the larger its death-peak." Phillips went on to show that regular people, not just the famous, seem able to postpone their deaths for special occasions like religious holidays and even national elections. Examining death records in New York and Budapest, Hungary, Phillips found a dip before and then a peak after Yom Kippur, the holiest and most important day of the year on the Jewish calendar. He found the same pattern before and after US presidential elections going back to 1904. After considering and rejecting many different theories, he concluded: "People postpone death in order to participate in social ceremonies. Because they are so attached to society, they die *post-maturely.*"

A professor of sociology at the University of California, San Diego, Phillips is a wonderfully energetic and engaging man who sprinkles conversation with references to William Shakespeare and various Amish, Arab, and Chinese proverbs. He's sixty-five years old, wears Birkenstock sandals every day of his life except to funerals, and enjoys rereading Agatha Christie and other detective novels in German so that there's some redeeming intellectual value for his efforts. Born in South Africa and educated in America, Phillips has devoted his entire career to the social and psychological factors affecting why and when we die. He's also the first social scientist to explore whether there is any statistical proof of the will to live.

Phillips found a treasure trove in California's death certificates. In one study, he analyzed 2,745,149 records from 1969 to 1990 and concluded that "women are more likely to die in the week following their birthdays than in any other week of the year." Specifically, Phillips found 3 percent more female deaths than expected in the week after their birthdays and a slight decline in the week before. Men's deaths peaked *before* their birthdays and did not rise afterward. Overall, he concluded, "people are more likely to die in the week after their birthdays than in any other week of the year."

Why would your birthday influence the timing of your death? Phillips argued the "best available explanation" was that women are "able to prolong life briefly until they have reached a positive, symbolically meaningful occasion. Thus, the birthday seems to function as a 'lifeline' for some females. In contrast, male mortality peaks shortly before the birthday, suggesting that the birthday functions as a 'deadline' for males." In other words, if you really dread all those candles on your cake, your chances of dying increase.

Over the years, Phillips's findings have generated considerable controversy. Indeed, an army of researchers has attempted to rebut most of his work. For instance, Dr. Donn C. Young at Ohio State University's Comprehensive Cancer Center analyzed a huge database of patients to see if there was a dip or peak in mortality around important holidays or birthdays. Examining 1.2 million deaths in Ohio from 1989 to 2000, including 309,221 in which cancer was the leading cause of death, Dr. Young concluded: "We find no evidence that cancer patients are able to postpone their death to survive Christmas, Thanksgiving, or their own birthdays."

I ask David Phillips about these studies that dispute his findings. He scoffs, saying that the only way to get published in a leading journal is to challenge someone else's work. I press him again. Can people really determine or postpone the timing of their deaths? Phillips says the will to live is an important crutch for many people. "No one wants to believe that our lives are buffeted by chance," he explains. "You want to believe that by trying harder you can live longer. It's comforting." After three decades of studying death certificates, he concedes that the anniversary effect is "very tiny" for most of us and perhaps a little more pronounced for famous people. While it's not *the* most important reason for survival, Phillips will wager a significant amount of money that the will to live is a contributing factor. "There isn't a single variable that explains everything," he says.

Phillips is absolutely correct. Admission to the Survivors Club occurs for many reasons. Some get in with their abiding faith. Others join with an abundance of luck. And many enter through sheer force of will. Alas, the fighting spirit can't prevail in every battle. But day after day, willpower makes a critical difference in who lives and who dies.

Survival Secrets

How Long Will You Live?

Jeanne Louise Calment lived 122 years and 164 days, the Guinness world record for the longest life ever. When she was born in Arles, France, in 1875, President Ulysses S. Grant resided in the White House, Jesse James was robbing trains in Missouri, Alexander Graham Bell was transmitting the twanging sound of a clock spring over a wire, and Leo Tolstoy was publishing *Anna Karenina.* Back then, life expectancy in France was under fifty years. Madame Calment took up fencing at age 85, rode her bike at

100, and lived on her own until she was almost 110. At 119, she gave up smoking two cigarettes a day, not for medical reasons but because she couldn't see well enough to light a match. At 121, she released CDs with her musings set to rap, techno, and regional music.

"I dream, I think, I go over my life, I never get bored," Madame Calment announced at her 121st birthday party in 1996. She died the following year. Many believe her longevity was inherited from her father who passed away at ninety-three and her mother at eighty-six, but good genes didn't help Madame Calment's only daughter, Yvonne, who died at age thirty-six from pneumonia.

The world's foremost authority on how we age is Dr. James Vaupel, a boisterous sixty-two-year-old who believes he's going to live at a minimum another twenty-eight years. He estimates that he'll die somewhere between the ages of ninety and a hundred. This prediction isn't wishful thinking. It's a forecast grounded in science and thirty years of studying longevity. Raised in New York and trained at Harvard, Vaupel is the founding director of the prestigious Max Planck Institute for Demographic Research in Rostock, Germany, where he presides over the Laboratory of Survival and Longevity.

The first question for Vaupel: How long can the rest of us expect to survive? In the big picture, he says, the news is very encouraging. At the moment, life expectancy around the world is growing. On average if you're an American, for every day that goes by, you can add five hours to your life expectancy. If you live anyplace else in the world, you get an additional six hours because life expectancy is growing even faster overseas. A few hours a day may not sound like much, but Vaupel says they really add up. For every six months that go by, you can tack on an additional five weeks to your life expectancy. Over the course of a decade, you get an additional two and a half years. Another way to look at it is to look backward and forward a hundred years. Life expectancy in 1909 was twenty-five years shorter than today and in 2109, people can expect to live twenty-five years longer, well past the age of a hundred.

That's all nice, but what about right here and now? What exactly determines how long you live? Most people believe their parents have a lot to do with it. Didn't Jeanne Calment survive 122 years because her mother and father lived so long? Not really, Vaupel says. Just 3 percent of your longevity is determined by how long your father lives and just 3 percent by your mother. By contrast, he says, 80 to 90 percent of your height is determined by how tall your parents are. All in all, your genes only influence about 25 percent of your longevity. "Genetic factors are not so important," he explains. "Even identical twins usually die many years apart." Indeed, studies of twins show that only 35 percent of their longevity can be explained by their identical genes. The rest—65 percent of their longevity—is up to their individual choices and life events.

So what matters most? Ten percent of your life span is determined in the womb when your mother's habits like smoking and drinking make a difference, and in childhood when your diet and exercise begin to influence your life. The rest—roughly two-thirds—comes down to the decisions you make and the things that happen to you, especially right now. "Are you taking good care of yourself?" Vaupel asks. "Are you currently eating well?" It's never too late, he says, to make changes that prolong your life.

Obviously, your life expectancy depends on many other

complex variables, including where you live. The people of Andorra, the tiny mountain principality wedged between France and Spain, have the longest life spans in the world, averaging 83.53 years at birth. As a group, the women of Japan are the longest-living people in the world, averaging eighty-six years. The longest surviving men are from San Marino, a twenty-three-square-mile republic surrounded by Italy, where life expectancy is eighty years. Of 223 countries ranked in terms of life expectancy, the United States comes in forty-seventh with an average of 78.14 years. By this measure, the worst place on earth is Swaziland in southern Africa, with an average of 31.99 years.

AVERAGE LIFE EXPECTANCY AT BIRTH (IN YEARS)

Top Five Countries:

1. Andorra	83.53
2. Macau	82.35
3. Japan	82.07
4. Singapore	81.89
5. San Marino	81.88

Bottom Five Countries (with world ranking):

1. Swaziland	31.99	(#223)
2. Angola	37.92	(#222)
3. Zambia	38.59	(#221)
4. Zimbabwe	39.73	(#220)
5. Lesotho	40.17	(#219)

Other Countries (with world ranking):

Canada	81.16	(#7)
France	80.87	(#8)
Sweden	80.74	(#9)
Australia	80.73	(#11)
Israel	80.61	(#13)
Italy	80.07	(#18)
Spain	79.92	(#22)
Germany	79.10	(#33)
United Kingdom	78.85	(#37)
United States	78.14	(#47)
Ireland	78.07	(#49)
South Korea	77.42	(#54)
Mexico	75.84	(#72)
China	73.18	(#103)
Brazil	72.51	(#114)
India	69.25	(#143)

SOURCE: *The CIA World Factbook*, 2008

So, to the extent you have any control, what tips the scales? Vaupel says loving life goes a long way. "People who live a long time are generally interested in life," he explains, pointing to Madame Calment, who enjoyed tennis, swimming, roller skating, and the piano. Vaupel also says, "It's not a good idea to be too tall. It's not a good idea to be too short. You should be somewhere in the middle." The ideal height is just over six feet. But don't worry if you're longitudinally challenged. There are exceptions to every rule.

If you want to prolong your life, what should you do right now? "What an individual can do is quite modest," Vaupel says. Not surprisingly, he recommends moderate exercise, limited saturated fat, wearing a seat belt, and installing smoke detectors. Sounds simple enough. But it gets even easier. Vaupel's favorite advice? "What people can do if they want to live longer is to listen to what their mother told them. We don't know much more than that." He runs down a list of reminders like, "Wear a hat when it's raining and a coat when it's cold." Don't forget to get eight hours of sleep, eat your fruits and vegetables. And there's one more: "Smile." It's the same simple wisdom that Madame Calment often shared with visitors. "Always keep your smile," she once said. "That's how I explain my long life. I think I will die laughing."

11

The Resilience Gene

Who Bounces Back and Who Doesn't?

One moment, she was carefree and cruising an Oklahoma highway in the middle of the night, speeding away on summer vacation. In the next instant, a jar of sulfuric acid came crashing through her windshield in a cruel explosion of glass and liquid fire. At first, the unbearable burning felt like a heart attack, but then Cindi Broaddus realized it was even worse. Her skin and clothes were melting from her body and dissolving in her hands. The smoldering acid would devour 75 percent of her face, and paramedics would dig the brassiere from her disintegrating chest. "In the blink of an eye," she says, "your life is never the same."

Cindi meets me for dinner at a Mexican restaurant off Highway 81 in Duncan, Oklahoma, a town that once boasted it was the Buckle on the Oil Belt. In this patch of America where a company called Halliburton was born in 1919, drilling rigs light up the darkness. At the restaurant, everyone greets Cindi with warm hellos. She's quite a celebrity around here because of the acid attack and because her brother-in-law is the psychologist and TV star Dr. Phil McGraw. Cindi looks great and takes care of herself, with blond highlights in her hair and a shiny red pedicure. She tells me she has gone through fifteen surgeries to rebuild her face. They repaired her eyelids, nose, and cheeks and grafted skin from her stomach to reconstruct her upper lip and chin. As soon as she starts talking, her green eyes sparkle and you don't really notice the scars. Cindi isn't a mathematician or a gambler but she figures the odds of what happened that night were one in ten trillion. "It's like I died and came back to life," she says.

Cindi's tale begins when the alarm clock went off around two in the morning on June 5, 2001. A single mother with three grown daughters, she awoke with excitement about her vacation in Southern California. She got dressed in a special outfit, carefully selected during a shopping expedition with her daughter Shelli. The new clothes looked a little trendy, especially the red capri pants. For a moment, she considered ditching her new stubby tennis shoes for open-toed sandals, but then she pictured Shelli's disappointment. She lived for her girls and just wanted to make them happy. As she laced up the chunky sneakers, she had no way of knowing this simple choice would prove critical in the ordeal ahead. So would her unbreakable devotion to her daughters.

Around 2:45 AM, Cindi hopped into the passenger seat of her black Sunbird. Her friend Jim Maxwell steered onto the H. E. Bailey Turnpike for the ninety-minute drive to the airport in Oklahoma City. They really needed this getaway. After all, Cindi rarely took time off, especially in the thirteen years since her husband had walked out on her and the girls. In those early days, with no job, experience, or skills, she struggled to pay the bills and feed her family. She suffered through some very dark days—she had cried her eyes out so many times after the girls had gone to school—but she had pulled her life together. She found a sales job at a cable company. A new man won her heart. Everything was looking up.

On this happy-go-lucky night, they were making good time on the highway that runs over the plains of southwestern

Oklahoma. The windows were open, and a warm breeze was blowing. Jim noted that the sky was unusually black—the stars and moon seemed to be hiding. Cindi bragged about her granddaughter Kennedy's first T-ball game. They also talked about the five-day trip ahead. She was so excited to soak up the San Diego sun, explore the famous zoo, catch a Padres game, and eat as much fresh seafood as she could handle. Around 3:30 AM, just outside Newcastle, the halfway point on their drive, Cindi decided to sneak a ten-minute power nap. The highway was empty, and Jim would be fine at the wheel. "Don't worry," he told her. "Your snoring is enough to keep me as alert as I need to be." After closing her eyes only for a moment, she suddenly felt a ferocious blast of pain.

"I can't see!" she screamed. "I cannot see. You've got to pull over. You've got to pull over. I'm dying. I'm dying."

"I cannot pull the car over," Jim answered. "I have to get us help."

Cindi was blinded by shards of glass, and her entire body felt like it was on fire. Many years earlier, her mother had passed away suddenly of a heart attack. The loss had been crushing. In those first moments of terror, she could only think of her daughters and the pain they would feel. "You have to tell my girls that I love them," she shouted into the wind that felt like it was blowing right through her.

"Stop it," Jim yelled back. "You're not dying. I'm not going to let you die. I'm going to get us help. And you'll be okay." Jim had been injured, too, burned on his head, arm, shoulder, and stomach, but the wounds were superficial compared to Cindi's. She kept begging him to stop the car. She just wanted the agony to end. In the midst of her pleading, she heard Jim's terrifying words: "We've been burned by acid. I can't stop."

Oh my God, she thought, *someone is trying to kill us.*

Now fear and pain attacked in tandem. She heard Jim say something about the lights of an Indian gaming center up ahead.

She could feel the car swerve to the right off the interstate. It was a curve in the road she remembers vividly. *Help is just right there,* she thought. *It's right there.* The tires screeched on the pavement, and Jim jumped from the car. "I'll be right back," he said. Without her vision, her other senses seemed more engaged and acute. She could hear Jim's footsteps in the parking lot. Then she heard his first word when he charged into the convenience store right before the door closed: "Call . . ."

Paramedics would soon arrive and rush Cindi to the burn unit at Baptist Medical Center in Oklahoma City. In the ambulance, they would search her entire body for a place to insert an IV needle in order to give her painkillers. Protected by those hefty new sneakers, her foot was the only spot that wasn't injured. The morphine would bring the first wave of relief.

To this day, the crime remains unsolved and the case is officially closed because the statute of limitations has expired. Cindi forgives the attacker, most likely a troubled male teenager in the area. She doesn't feel hate. She doesn't want revenge. She helped pass a new state law that makes it a felony to throw objects that inflict injury or damage from bridges and overpasses. For justice, she only wishes the mystery man who hurt her could really understand the agony he inflicted. To know the truth, she says, he needs to spend a few months in a burn ward seeing all the wounds that won't heal and hearing the cries that won't end.

It's getting late and we're finishing dinner. Cindi says she's just an ordinary woman who has been forced to handle an extraordinary amount of adversity: agonizing loss and grief, failed relationships, economic struggle, and unbelievable violence. Somehow each time, she finds a way to bounce back. "You can ask a million people what they would do if a gallon of acid was dropped on them," she says, "but until it happens to you, you don't know—you really don't know how you would react and what you would do." She stops to take a sip of her ice water. "If someone says, 'Hey, do you want someone to throw acid on you? Do you want to live your life totally scarred on the outside? And people staring at you? Do you want to go through fifteen surgeries and go through the horrible, horrible experience of skin grafts?' I'd say, 'Hell no!' But would you walk this journey again to come out on the other side the person you are today and to have experienced the love and kindness of strangers and family? I would do it in a heartbeat. I would walk this journey again in a heartbeat."

1. The Long and Short of Resilience

Very few people are as resilient as Cindi Broaddus. Lots of folks slip into a funk or depression after a relationship breaks up or a terrible medical diagnosis gets delivered. Many binge on alcohol and drugs. Still others lash out in anger and violence. You probably know people who can't seem to handle life's bumps and bruises. When they're knocked down, they struggle to get up. You also probably know a few people who seem impervious, almost invincible. When they run into hard times, they rebound almost immediately. What accounts for the wide range of behavior? How did Cindi pull herself together after the death of her parents, the breakup of her marriage, and the acid through the window? Why didn't she spiral out of control? A person's psychology and life experience definitely play important roles, but what if your response is also influenced by microscopic proteins that you inherited from your parents and that were programmed from birth? In short, what if there's a Resilience Gene?

It sounds extraordinary, but it's reality. Just ask Dr. Richie Poulton, a professor of preventive and social medicine at the University of Otago in New Zealand, who runs the most important ongoing study of human health and development ever undertaken in the world. In 1972, scientists began to study 1,037 babies born in Dunedin, a lovely harbor town on the country's South Island. For the past thirty-six years, the Dunedin study members—as they're known—have been meticulously tracked. When they sprain an ankle or break a bone, the information is carefully collected. When they get married or divorced, it's painstakingly recorded. When they struggle with a bank payment or get fired from a job, it's duly chronicled.

Working with an international team, Poulton analyzed the stresses that the Dunedin members encountered between the ages of twenty-one and twenty-six.* Specifically, he looked at fourteen different kinds of everyday challenges ranging from unemployment and mortgage troubles to divorce and domestic abuse to chronic disease and injury. Next, the researchers studied the DNA of each Dunedin member. The goal was to define and quantify the relationship and interplay between their genes and life stresses. In science shorthand, it's called G x E or genes times environment. Incredibly, they discovered that one specific gene significantly influences how we respond to stress. People with the so-called protective version of the gene are much more resilient, almost immune, to life's adversities, while those with the stress-sensitive version are at greater risk of depression and suicidal thoughts. The findings were truly groundbreaking. It was the first time scientists had studied the complex interactions among people's genes, the adversity they faced in life, and their risk of depression. Even more significant, they identified the specific version of a gene—or genotype—that seemed either to protect people from these problems or to put them at higher risk.

The test for the Resilience Gene isn't commercially available, but I'm very curious to find out if I've got it. So I ask the DNA experts for a little inside help. Not long after, a "Mouth Swab

---

* The team was led by psychologists Avshalom Caspi and Terrie Moffitt of King's College, London, and the University of Wisconsin–Madison.

Collection Kit" kit arrives from England in a small padded envelope. The package contains a red-capped test tube with clear preservative solution inside and ten little wooden sticks with cotton tips. "Each bud should be rubbed on the inside of the cheek for approximately 20 seconds," the instructions say. I sweep the Q-tips around the inside of my mouth and then jam them into the plastic beaker. It takes a few minutes and I'm done. I wrap the test tube in a plastic bag, seal it in the return envelope, and send it off to the Institute of Psychiatry in southeast London, where it will be analyzed. In a few weeks, they'll send the results. Needless to say, I'm quite anxious to know if I have the Resilience Gene.

Professor Ian Craig runs the Molecular Genetics Section of the prestigious SGDP Center at the Institute of Psychiatry in London.* He was responsible for testing and genotyping the 847 Dunedin study members. More specifically, the sixty-five-year-old scientist looked for variations in each person's serotonin transporter gene—known as 5-HTT—which helps regulate serotonin in the central nervous system. Serotonin is the brain's feel-good neurotransmitter, a kind of self-esteem hormone that plays a critical role in managing anger, mood, sexuality, and appetite, among other things. Low levels of serotonin are often associated with depression, anxiety, and aggression. Professor Craig explains that the 5-HTT gene comes in two different sizes or alleles. There's a relatively short, inactive version that is stress-sensitive, and there's a longer, more active version that appears to protect against adversity. Each of us inherits two copies of the gene, one from our mother and one from our father. That means each of us has one of three possible combinations of the 5-HTT gene: two short versions, one short and one long, or two long. In their pioneering study, Craig and his colleagues found

* SGDP stands for Social Genetic and Developmental Psychology.

that your resilience to life's everyday stresses depends significantly on the version of the 5-HTT gene that you inherited. If you literally got the short end of the 5-HTT stick, you're more vulnerable to depression. More precisely, Craig notes, if you've got short, stress-sensitive alleles, you're two and a half times more likely to suffer depression after everyday life challenges than people with long, protective alleles.

You're probably wondering: *Which version do I have?—short or long?* Sorry to say but you're probably at some risk. Here's why: 17 percent of us have short, vulnerable, stress-sensitive alleles. That's the most at-risk group. Another 51 percent of us have a mixture of one short and one long allele. This large, middle category isn't as vulnerable as the first group, but members are still 1.7 times more likely to suffer depression after tough life events than the most protected people. In total, 68 percent of us are at medium or high risk of depression after stressful life events because of our short or mixed 5-HTT genes.

That leaves 32 percent of us with long, active, protective alleles. In other words, only one-third of us have the Resilience Gene. That doesn't mean we're invulnerable to life's stresses. It just means that we're wired better to rebound from the problems everyone faces. In the Dunedin study, only 17 percent of people with the Resilience Gene developed depression after experiencing various stresses like a death in the family, losing a job, or breaking up with a partner. No matter how many tough life events they encountered, these people were at *no greater risk* of depression than those who had never encountered any stressful events at all. It's almost like the bad stuff never happened to them.

How precise is all of this? Craig emphasizes that one gene alone doesn't determine your response to life's toughest challenges, especially when you consider that you have twenty-five thousand other genes, and many different environmental and psychological factors come into play. "The predictive power of it is very slight," Craig says. That's why he and his colleagues aren't ready to use 5-HTT for diagnosing your level of resilience or vulnerability. For now, the discoveries about 5-HTT offer a glimpse of what's to come. Someday scientists will be able to identify all of the genes that influence your response to stress and design medicines to protect you from life's inescapable hardships.

A couple of weeks after I sent my Q-tips to London, I receive an e-mail with the subject heading "DNA genotyping." I hold my breath as I click on the message and scan the results. Alas, it turns out that I don't have the Resilience Gene. I'm in the big middle group of people with some protection against life's stresses but not the most. In other words, I'm somewhat resilient and somewhat not. When times are tough, I'm at some risk. I try to take solace from the fact that it could be worse: I could be in the group with two short alleles and no protection. Like everyone else, I'll have to wait for the day when they make pills to boost resilience. But I'm still curious about people like Cindi Broaddus who always bounce back from the worst. Besides their genes, what else keeps them going when others surrender or give up? The question leads me to New York City to see an old friend who works at the very top of Rockefeller Plaza.

2. "A Real Kick in the Pants"

Jeff Zucker relaxes in a cream-colored armchair on the fifty-second floor of 30 Rock. He's literally on top of the world. It's been a big week for the forty-one-year-old television executive. Just three days earlier, he was appointed president and chief executive of NBC Universal, capping an extraordinary, almost gravity-defying rise in the television business. Zucker is a human whirlwind, but it's surprisingly calm and quiet in his office. I've known him since the early 1980s when he was a skinny sports reporter on *The Harvard Crimson*. Back then, he had a huge curly head of hair, but most of that is gone now. Much is made of his size—he claims he's five foot six on a good day—but no one ever thinks of him as diminutive. At the preposterously youthful age of twenty-six, Zucker took charge of NBC's *Today* program, leading it to an unprecedented streak of dominance and influence in morning television. In the control room, where dozens of TV screens flash images from cameras and satellite feeds around the world, Zucker was a genius under pressure. I may be biased because we worked together at NBC News, but many of our colleagues agree that he possessed an extraordinary ability to stay focused and see clearly no matter how much chaos surrounded him. The greater the pressure, the more decisive and determined he grew. The tougher the times, the more confident and optimistic he seemed. Sure, a volcanic temper sometimes erupted, but his fierce competitiveness inspired incredible loyalty and led to astonishing success.

I've come to see him today for a simple reason: Zucker is the most resilient person I've ever known. If anyone has two long, active 5-HTT alleles, he does. He's survived more than two decades in the kill-or-be-killed world of network television. At the *Today* show, he endured the constant stress of producing live television every morning, not to mention the brutal 4 AM alarm clock that wakes you up for work. Later, just after his four-year stint as president of NBC Entertainment, the network plummeted from first to last place in the ratings. And in the midst of it all, he fought two bouts with colon cancer. The man seems both indestructible and unstoppable.

I show him the front page of *The New York Times* business section with the latest profile chronicling his relentless rise. The headline reads "NBC's Own Survivor." In the next line, the *Times* declares: "It wasn't just Jeff Zucker's many successes that took him to the top of NBC Universal. It was his ability to handle the tough times." He smiles and shrugs. It turns out he doesn't ever think about being a survivor. "I never dwell on that fact," he says. Zucker admits he isn't the most introspective person, but when it comes to his cancer, he's done his share of ruminating—and crying—about life and death. He clearly remembers a certain Tuesday night in October 1996. He had only been married five months and was on his way out the door with his wife, Caryn, to see the off-Broadway play *I Love You, You're Perfect, Now Change.* The phone rang. It was a doctor who had run some tests because Zucker hadn't been able to move his bowels for ten days and felt an odd tingling sensation in his stomach. The doctor said "he had a real 'kick in the pants' for me," Zucker remembers. He slowly repeats the phrase *kick in the pants*. The physician informed the thirty-one-year-old producer that he had colon cancer. "I couldn't believe it," Zucker says. "I sat down on the couch and I cried with Caryn."

Three vivid memories come next. "The first is being wheeled into the operating room the first time," he says. It was freezing cold, and a nurse stroked his hand to relax him. It was Halloween 1996. He remembers the date because while he waited for surgery, he watched the *Today* show's costume contest, an event that he had created, with Katie Couric dressed up as Marie Antoinette and Bryant Gumbel as Jimi Hendrix. Next, Zucker says, "I remember vividly waking up in the recovery room." After some twelve hours of surgery, he experienced "unbelievably intense pain up and down my stomach where I had been ripped open and stapled back together." Zucker asked his wife, "Am I okay?" She answered, "Yes, they got it all." He grimaces about the third memory: the awful metallic taste of the chemotherapy that he received every Friday after producing the show.

Each time there was a setback in his cancer battle, Zucker rapidly absorbed the blow and focused on what needed to be done. When doctors came to his hospital room after the first surgery to announce the cancer had spread to his lymph nodes, Zucker replied: "'Okay, what do I have to do? Let's just do it. Because the sooner we get it over with, the sooner I'm going to be better.' And that is how I attacked it and approached it."

How does he explain his resilience? "Beyond the biology," he says, "I think I am very driven to succeed. I am not saying that I willed myself to wellness, but I certainly did not want to let my illness get me down or define me." From an early age in Miami, Zucker was an incredibly competitive tennis player. "I always hated to lose," he says. "I always placed that pressure on myself." He stops for a moment to reflect. "I don't know why. I don't know if it was because I was usually the smallest kid in the class or the smallest kid on the tennis court and I felt I had to prove myself. I don't know. That's a little psychological perhaps. Maybe I felt the need to overcome that, stand out a little, I don't know."

A devoted father of four young children, Zucker is unabashed about his love of winning. Whether it's television, tennis, or cancer, he wants to dominate. Zucker's boss is Jeffrey Immelt, the chairman of General Electric, which owns NBC-Universal and has more than three hundred thousand employees around the world. Even at the summit of corporate America, Zucker's aggressiveness stands out. "Jeff is the most competitive executive at GE," Immelt has said.

Zucker gestures toward the small blue-and-orange figurines that stand on the windowsill behind him. On closer inspection, they're miniature football players, replicas of the heroes of the undefeated Miami Dolphins of 1972. A tiny statue of Coach Don Shula stands alongside quarterback Bob Griese and running backs Larry Csonka and Mercury Morris. "That was an extraordinary season to become a football fan," Zucker says. His team went 17–0, winning the Super Bowl against the Washington Redskins. "When your team is like that right from the start," he asks, "don't you have an expectation that you're supposed to win?" The argument seems perfectly logical to him—if your team is undefeatable, aren't you supposed to be invincible, too? But then he turns the argument around. As well as anyone, he knows that the good times don't last forever. Winning streaks are broken. "Life is not a straight line up," says the man on the fifty-second floor of Rockefeller Plaza with a view all the way down the island of Manhattan. "Things are not always perfect. The measure of a person is how they deal with adversity."

3. The Resilience Prescription

In just a few chapters, you'll have the chance to take the Survivor IQ test and discover if resilience is one of your leading strengths. Even if it isn't, don't despair. You can boost your ability to bounce back with a few easy steps. Dr. Andy Morgan of the Yale School of Medicine has spent many years studying posttraumatic stress and the brain chemistry of highly resilient US Special Forces. He believes you might want to make a few changes in your diet during stressful periods in order to protect your brain and body and recover faster. The first place to start: carbohydrates. Dr. Morgan studied three groups of soldiers at Fort Bragg, North Carolina, who had undergone intense stress in survival training. Along with identical meals, they were given Kool-Aid drinks with different amounts of maltodextrin, an easily digestible sugar, or beverages with no sweeetener at all. The next morning, the soldiers were given a series of physical and problem-solving challenges. It turns out the carbs helped the troops recover faster and get back to work. Indeed, the soldiers with sugar in their Kool-Aid were able to hike faster and set up ambush sites quicker. Dr. Morgan also found that the carbs made people smarter and faster on cognitive tests using parts of the brain that govern attention and concentration. For the military, these findings mean that medics may someday pump soldiers full of maltodextrin or other carbs to accelerate their recovery from stress and get them back into the fight. For the rest of us, Dr. Morgan says, it may make sense to load up on carbs after a grueling day in order to rebound faster. "That's kinda cool," Dr. Morgan says with enthusiasm. Carbs are cheap, they're easy to consume, and there aren't any significant side effects except if you overdo it and start gaining weight. Carbs also trigger the release of serotonin in the brain, which makes you feel better. To enhance your resilience in the long run, Dr. Morgan says, you might also consider consuming more DHA, an omega-3 fatty acid that can be found in fish, eggs, certain kinds of meat, and supplements. Various studies have shown diets rich in DHA—docosahexaenoic acid—can help people with depression and perhaps prevent the development of Alzheimer's later in life.

Dr. Dennis Charney is dean of the Mount Sinai School of Medicine in New York. The ebullient fifty-seven-year-old is the reigning king of resilience studies in America. He has taught, trained, or worked closely with most of the country's experts in the field. When I heard that he had developed something called the Resilience Prescription, I went to see him in his office on the twenty-first floor of a high-rise on Manhattan's Upper West Side. Dr. Charney invites me to his conference table. Regardless of your genes or where you're from, Dr. Charney tells me, everybody can boost his or her resilience by following a ten-step plan that he developed with Dr. Steven Southwick, a professor of psychiatry at the Yale School of Medicine. As you'll see, many of their recommendations touch on themes in this book.

The first and most important part of the Resilience Prescription, Dr. Charney explains, is to practice optimism: "That's a biggie." Some people are born optimistic; others learn it. Either way, studies show that optimists respond better when they're diagnosed with breast cancer and they experience fewer problems requiring hospitalization after heart surgery, Dr. Charney says. Of course, as we saw in chapter 2 with the Stockdale Paradox, unguarded optimism can make your problems even worse. The key is to stay positive and hopeful while confronting reality. The second step in Dr. Charney's prescription is to identify a resilient role model. You can choose someone you know or someone who inspires you. Lance Armstrong is a perfect example. So is your aunt Betty who has beaten breast cancer. Imitation is an excellent way to learn to be more resilient, Dr. Charney says. Third, develop a moral compass and unbreakable beliefs. Resilient people find strength in God, and in ideals and principles greater than themselves. Fourth, practice altruism. By helping others, Dr. Charney says, you can help yourself feel better in tough times. Fifth, develop acceptance and cognitive flexibility, meaning the ability to learn and adapt your knowledge and thinking to new situations. Sixth, face your fears and learn to control negative emotions. Seventh, build active coping skills to handle your problems. Eighth, establish a supportive social network to help you. Ninth, stay physically fit. And tenth, laugh as much as you can.

It's an ambitious list, and Dr. Charney intends to publish his formula some day in a book. You don't need to try all ten steps at the same time, he tells me. Focus only on a few and you'll see results. And even if you don't have the long versions of the 5-HTT gene, he insists you can still improve your resilience. Leaning forward with his elbows on the table, he observes that the public is mesmerized by the dark side of human psychology and aberrant behavior. As an example, he cites Cho Seung Hui, the shooter at Virginia Tech who massacred thirty-two students and faculty in April 2007. As alarming details about the twenty-three-year-old killer began to surface, Dr. Charney says, many people wondered: "How do you make a human that bad?" He speculates that genes played a part in the murder spree. So did environmental factors, including the shooter's childhood in South Korea and upbringing in the United States.

But while it's intriguing to probe the criminal mind, Dr. Charney believes we often overlook something more powerful and universal. The greatest surprise of his career, he says is the "hidden capacity" of most people to rebound from adversity. As we have seen, this is the third rule of the Survivors Club: You're stronger than you know. Dr. Charney literally overflows with astonishing stories about prisoners of war, violent-crime victims, and members of his own staff who have recovered from real trauma. In the world of medicine—and certainly the media—it's tempting to focus on evil and aberrant behavior. But if you look at all the resilience around you, Dr. Charney says, you can't help exclaiming: "Wow, being a human can be a beautiful thing."

Survival Secrets

Are Your Initials Killing You?

My initials are BS, and I've endured a fair share of teasing about them over the years. Early on, I learned to conceal them by adding my middle initial B. Presto, no more BS. The scatological insults disappeared. It turns out BBS is totally innocuous. If you consider the controversial Theory of Killer Initials, BBS may also be far less deadly.

If you've got initials like ASS or PIG—letters that spell words with negative undertones—Professor Nicholas Christenfeld has some grim news for you. Bad initials kill. A professor of psychology at the University of California, San Diego, Christenfeld focuses on how tiny little irritants—called stressors—can add up over the course of a lifetime and contribute to health problems and even premature death.

In 1999, Christenfeld and two colleagues analyzed the names of people who died in California from 1969 to 1995. They came up with lists of initials that spelled out positive and negative words. Good initials included ACE, WIN, WOW, and VIP. Bad initials included RAT, BUM, SAD, and DUD. Then, they examined whether people's initials had anything to do with how long they lived. All told, they studied around thirty-five hundred men and women with good and bad initials.

The results were stunning. As far-fetched as it may sound, Christenfeld concluded that your initials can actually influence the time *and* cause of your death. "A symbol as simple as one's initials can add 4 years to life, or subtract 3 years from it," he wrote in a paper published in the *Journal of Psychosomatic Research,* a well-regarded academic publication. More precisely, men with positive initials like GOD, HUG, and JOY lived 4.48 years longer than those with neutral initials. Men with negative initials like HOG, BAD, APE, and DIE lived 2.8 fewer years. For women, the effects were smaller: Positive initials meant 3.36 more years of life while negative initials had no impact. What follows on the next page are the initials that Christenfeld studied:

POSITIVE INITIALS	NEGATIVE INITIALS
ACE	APE
GOD	ASS
HUG	BAD
JOY	BUG
LIF	BUM
LIV	DED
LOV	DIE
LUV	DTH
VIP	DUD
WEL	HOG
WIN	ILL
WOW	MAD
	PIG
	RAT
	ROT
	SAD
	SIC
	SIK
	UGH

Intrigued but dubious, I called Christenfeld, an engaging and persuasive forty-five-year-old from La Jolla, California, to ask directly: How do bad initials like MAD and BUG actually kill? Cumulatively over a lifetime, Christenfeld replies, they build up as "a little tiny source of stress" or they "crush slightly your will to live in some way." The impact isn't immediate, he says. Instead, like compound interest, it accumulates over years and years, including every time the bank teller asks you to initial a check and you have to write ZIT or FAT. "Every now and then, you have a little tiny depression in your health or your immune system or your self-esteem," he explains. "It's a little tiny stressor for you but it happens again and again and again, decade after decade." Christenfeld is quick to note that people who initial SIN, OAF, or NAG on their rental car papers don't just drop dead on the spot. Instead, "It is the kind of thing that can contribute slightly to the total stress on the system," he says. Your initials alone don't finish you off. Rather, they contribute to all the other damage we do to ourselves over a lifetime, like not eating enough roughage, smoking cigarettes, drinking too much, or getting yelled at by our boss every day. On the flip side, good initials "are some source of tiny pride or self-esteem." Seeing the monogram FOX, GOD, or HUG on your key chain or robe every day can add up to something positive and meaningful over a lifetime.

Christenfeld admits that he was "highly suspicious" when he first discovered the results. He and his colleagues worked hard to find flaws in their thinking. Obviously, they wanted their research to be accurate, but they also didn't want to humiliate themselves with findings that would be ridiculed. Each time they pored over the data, they grew more and more convinced. Indeed, when Christenfeld dug deeper, he found yet another surprise. People with positive initials died less frequently from causes that often have psychological or behavioral explanations like suicide and accidents. Presumably, he theorizes, they were happier and took fewer risks—like drinking and driving—that commonly lead to accidents. Indeed, people with positive initials experienced 33.7 percent fewer accidental deaths than people with neutral initials. On the other hand, people with bad initials died more frequently from causes like suicide and accidents.

The Theory of Killer Initials has generated a fair amount of controversy. In a paper titled "Monogrammic Determinism?" two economics professors at Pomona College in Southern California examined the same data and concluded that initials have no effect on life and death. Now Christenfeld and his colleagues are rebutting the rebuttal. In the end, he admits he wouldn't stake his life on this research. But he also believes people should give a little extra thought to the selection of names.

I ask about his own initials, which are NJSC. "Nothing interesting at all," he replies, except that he's got four instead of three. "The vast, vast majority of people have initials that just don't mean anything, and they're neither a source of pride nor despondency. They're just gibberish." I inquire about his two children. Did he pick positive initials to give them an extra boost in life? "I didn't know about this when I named them," he says, adding it's quite a relief that they didn't end up with killer initials like SIC from his study.

12

What Does Not Kill Me

Why Adversity Is Good for You

In the Psychology Department of the University of North Carolina at Charlotte, you'll find office door #4027 with a cartoon pasted on it. The illustration shows a doctor in a white coat falling from a building that is named the INSTITUTE FOR THE STUDY OF EMOTIONAL STRESS. It appears that he has been thrown out a window, and the caption quotes a patient inside who says: "I feel better already." Welcome to the irreverent and surprising world of Dr. Lawrence Calhoun. In the jargon of his field, he's an extrovert: a gregarious and funny man with a gift for telling stories. He moved in his teens to North Carolina from Brazil, but his accent is entirely Southern. You might say that Dr. Calhoun's office is the global headquarters of a relatively new and surprising field of psychology. The shelves in libraries are filled with volumes on posttraumatic stress, and the public is constantly exposed to news about the ravages of PTSD. But Dr. Calhoun and his colleague Dr. Richard Tedeschi are exploring the unexpected flip side. It's a little-known phenomenon that they named posttraumatic *growth*.

"Most people exposed to the worst traumas do *not* experience psychiatric disorders," Dr. Calhoun tells me. "They turn out fine." Contrary to what you might expect, serious psychiatric problems are not a necessary and inevitable consequence of life's worst crises. In fact, Dr. Calhoun believes that most people actually are transformed for the better from their battles with life's toughest stuff. While this phenomenon draws scant attention in the media, let alone academia, Dr. Calhoun argues that posttraumatic growth is significantly more prevalent than posttraumatic stress. Over the years, he has worked closely with hundreds of patients suffering from trauma, and his colleagues and students have studied survivors of every kind of ordeal, including violent crime, natural disasters, serious illness, divorce, unemployment, disability, grief, and loss. In almost all of their research, he encountered an "amazing paradox": Good things emerge from the worst experiences. It's certainly not an original idea, Calhoun admits. "What does not kill me makes me stronger," Nietzsche wrote famously, and sure enough, literature and religion teem with references to growth from adversity.

The concept may be familiar, Calhoun says, but its pervasiveness is truly surprising. "What we have found new and remarkable," he writes in *Trauma & Transformation,* "is how often [growth] happens and how apparently ordinary people achieve extraordinary wisdom through their struggle with circumstances that are initially aversive in the extreme." Dr. Calhoun cites studies showing that between 30 to 90 percent of persons facing serious crises experience at least some positive change.

In what other ways do people grow from trauma? A significant number of survivors report that their relationships are strengthened by their ordeals, says Dr. Calhoun. They experience greater compassion and sympathy. They feel simultaneously more vulnerable in the world—and stronger. Their philosophies of life improve with a shift in priorities as they recognize the precariousness and limits of their time on earth.

Dr. Calhoun doesn't discount the frequency or seriousness

of posttraumatic stress. Nor does he believe that growth and stress are mutually exclusive. Indeed, both can exist at the same time. The important point, he notes, is that trauma isn't purely negative and destructive: Growth through suffering is "an experience as old as humans."

What does posttraumatic growth really look like? To test the theory, let's briefly consider the case of an ordinary mom in an extraordinary situation. Cassi Moore's California license plate is ONETHMB. Her e-mail address is Onethumber. In June 1998, Cassi was pitching a tent on the Navarro River in Northern California when she cut her thumb. It was a small wound, no big deal, and she quickly forgot about it. A couple of days later, she started feeling achy. When she went to see her doctor, he sent her home telling her it was just the flu. Within a day, Cassi was so weak she couldn't lift her hands. This time, the doctor told her not to worry: It was food poisoning. Not long after, Cassi's left side suddenly started to turn black, oozing blood, and she began having trouble breathing and seeing. When she was rushed to the emergency room at Palm Drive Hospital in Sebastopol, her blood pressure was undetectable and her body was in septic shock, an extremely serious condition in the blood and tissues brought on by infection. Finally, doctors diagnosed the problem: necrotizing fasciitis, better known as flesh-eating bacteria. It had all started with that little cut on her thumb. They hurried Cassi into surgery and removed a large chunk of her left side and breast. Then she fell into a coma for twenty-two days.

When her eyes opened in the intensive care unit, Cassi listened quietly as her husband and parents tried to comfort her. Just thirty-two years old, she looked down and discovered that nine of her ten fingers were black and shriveled along with all of the toes on her left foot and her right foot all the way up to the ankle. Only her left thumb was still healthy. When they told her the frightening name of the disease, she burst into tears. She had seen a *20/20* episode on television about necrotizing fasciitis in England with victims whose faces had been ravaged by the disease. She knew it was a rare infection of the deep layers of the skin that was fatal around 30 percent of the time. "I just thanked God," Cassi says. "Thank you, God. Thank you, God. Over and over because I knew I should have been dead." She spent seventy-seven nights in the hospital but remembers only one really bad day, when doctors removed the bandages after amputating nine fingers. "I was very upset because I thought I would never play my keyboard and guitar again," she remembers. But the tears dried quickly and she focused on getting better. *You know what?* she thought. *It doesn't matter that I've lost this stuff. I'm here for my three kids.*

The bacteria destroyed 6 to 8 percent of her flesh, but Cassi believes it also forced her to become a better wife, mother, and human being. Two years before the crisis, her marriage and faith were on the rocks. But she believes those struggles were God's way of preparing her for what was ahead. "He sees the future," she says. "He knew this was coming." When they celebrated their twentieth wedding anniversary in Palm Springs, Cassi and her husband agreed they were actually grateful for the crisis. It gave them a chance to grow closer to God and to each other. Once a pessimist and worrier, Cassi now says she's an optimist. She's forty-one and more cheerful than ever. She plays her instruments, laughs all the time, and makes jokes with her kids about throwing her prosthetic leg out the window while driving around their hometown of Windsor. "I don't know that I'd trade it," she says of her ordeal. "Yeah, it was a horrible thing," she continues, and there's no denying that she has only one thumb. "But I'm always happy now." She pauses to reflect on how improbable those words must sound. "I'm not kidding," she goes on. "I'm always upbeat, I'm always smiling. I've always got a joke to tell." She hesitates again. But facts are facts. "It's true," she insists of her personal growth after trauma. "Ask anyone."

1. The Surprise of the POWs

On February 12, 1973, the first American prisoners of war were released from Vietnam and began the long journey back to the United States aboard an Air Force C-141 Starlifter nicknamed the Hanoi Taxi. The Pentagon had spent more than a year planning Operation Egress Recap, the sensitive mission to repatriate more than five hundred POWs who had survived torture, isolation, and extreme deprivation. From previous experience, returning POWs were known to suffer guilt and shame upon release, so this time the military provided them with tailored uniforms and appropriate medals and insignia of rank to wear for their homecoming. The aim was to offer a little comfort and to remind many that they had received promotions during their captivity.*

As the first POWs arrived home, a young navy doctor named Robert Mitchell was waiting to take care of them. Dr. Mitchell had helped organize Operation Egress Recap with special focus on the physical and mental health of the prisoners. Many had been seriously injured from abuse and beatings. Some were seriously ill from medical neglect. Others suffered from psychological trauma. But to his great surprise, Dr. Mitchell remembers, most of the men came home in remarkably good shape. And he marveled how quickly they rebounded from their physical and psychological wounds. Indeed, many seemed to grow, even benefit, from their ordeals. "Quite honestly," he tells me, "I didn't expect them to do as well as they did." With pride and awe, he rattles off their accomplishments over the years. At least seven

* The acronym *POW* is employed here in its common usage. More accurately, prisoners released from captivity are ex-POWs, former POWs, or repatriates.

earned flags for the rank of admiral in the navy, while others like John McCain, Jeremiah Denton, and Sam Johnson went on to serve in the US Senate or House of Representatives. The leader of the POWs, Vice Admiral James Stockdale, ran for vice president with Ross Perot in 1992, and many former prisoners are professional leaders in medicine, business, and law. "It's truly an outstanding group of men," he says. "I just love these people."

Dr. Mitchell—who retired from the navy in 1990 with the rank of captain—is something of a survivor himself. He's ninety years old now and moves with the help of a walker. He spends most of his time caring for his wife, who has Alzheimer's, but he still gets calls in the middle of the night from POWs and he still counsels their wives when they reach out for help. Despite his years, Dr. Mitchell remains the country's foremost expert on the physical and mental health of POWs. His mind is sharp and his eyes sparkle when he talks about the "fellas."

If you serve in the military and are captured as a prisoner of war, you're guaranteed a lifetime of free medical and psychological visits when you come home. You're sent to the Robert E. Mitchell Center for Prisoner of War Studies. It's located on the vast grounds of the naval air station in Pensacola, Florida, where the streets are lined with shady trees draped in Spanish moss. The Mitchell Center is housed in Building 3933, a modern redbrick structure, across from the Barrancas National Cemetery and its sweeping rows of white headstones. The center is a serene, quiet place with linoleum floors and walls decorated with surveillance photos of notorious POW camps named Alcatraz, Briarpatch, and Plantation. There are also framed snapshots of the POWs who come here almost every day from homes as far away as Greece and Thailand.

Some 570 POWs from Vietnam are still alive, and the center encourages them to come yearly for extensive checkups. Most do, and a handful of ex-prisoners from World War II and Korea also visit, along with twenty-one former POWs from the first Gulf War and a handful from the Iraq war. The center also sees State Department employees and civilian contractors who found themselves in the wrong place at the wrong time. Dr. Bob Hain is the director of the center. Warm and welcoming, he joined the navy in 1978 as a flight surgeon and, like everyone at the Mitchell Center, speaks of the POWs with awe and admiration. The "vast majority" of them, he tells me, have done "fairly well in life." He pauses and then corrects himself: "Quite well actually."

Dr. Hain and his staff keep careful track of every prisoner and refer to them on a first-name basis. Each detail of their lives is carefully tallied and analyzed. The results are then compared with a nearly identical group of Vietnam veterans who weren't imprisoned during the war. The goal is to understand every dimension of the POW experience, especially how they have fared since returning from Southeast Asia.

Dr. Jeff Moore, the neuropsychologist or "shrink" at the center, has worked with POWs for the last nineteen years. He believes the greatest mystery is "why most of them don't turn out to have chronic PTSD." In other words, why aren't more of them "psychiatrically ill"? Among navy and air force aviators in general, Dr. Moore explains, the incidence of PTSD is 18 percent. Among prisoners of war in North Vietnam, the rate isn't much higher—24 percent—a number that seems surprisingly small given the trauma they endured.* So why did they do so well in captivity and upon returning home? Dr. Moore believes the American aviators captured in North Vietnam were uniquely prepared for their ordeals. For starters, in order to fly missions over Vietnam with its concomittant chance of being captured, you were already a member of an elite, well-

* Among the American POWs in South Vietnam—where prisoners were typically enlisted men who did not have the benefit of specialized survival and resistance training before deployment—the PTSD incidence is 60 percent. Among American ground troops in Vietnam who were not POWs, the rate of PTSD is 30 percent.

prepared group. American pilots were rigorously screened for physical and psychological fitness. They were highly intelligent, competitive, resourceful, adaptable, and optimistic. They were trained to divide big problems into manageable tasks and attack their challenges head-on. They were experienced in handling tremendous stress and actively coping with adversity. They had enormous faith in themselves, their comrades, and their country. Indeed, Dr. Moore says, it's hard to imagine a group of men better equipped for the specific adversity they faced. Of course, they weren't impervious to Vietnamese interrogations and torture. At some point, he notes, every one of them broke—but almost all recovered. The twin creeds of the POWs were *Bounce Back* and *Return with Honor,* and they succeeded remarkably on both counts. In fact, most of them feel better and wiser *because* of their captivity. Imprisonment and torture deepened their commitment to God, country, and their fellow men. If the Vietnamese didn't kill them, Dr. Moore says, the experience made them stronger. One study actually quantified this phenomenon and found that 61 percent of POWs reported "favorable" psychological changes after captivity, including a deeper understanding of themselves and others and a clearer sense of life's priorities. Remarkably, in another study of American prisoners in the Korean War, one quarter reported that they learned so much from the experience that they "would be willing to go through it again."

I ask Dr. Moore what the POW experience can teach the rest of us. Can we extrapolate from their posttraumatic growth or are these men simply too different from regular people? Dr. Moore leans back in his chair. "Human beings aren't as vulnerable as some people think," he answers. Over the course of our lives, he goes on, between 70 to 75 percent of us experience major traumas and crises that are sufficient to trigger stress-related disorders. And yet, mysteriously, only 8 to 12 percent of us experience symptoms of PTSD. That means more than

60 percent of us experience trauma but don't suffer psychiatric disorders as a result. Dr. Moore grins at this revelation. "Most people are a lot more resilient than they realize."

I try the same questions on Dr. Mitchell, the physician who has spent more than thirty-five years taking care of POWs. Impossible as it sounds, he believes the majority of us "would do quite well" if we faced the same challenges as the prisoners. I press him on this point. Ordinary people aren't as physically fit as navy pilots. We aren't trained to resist interrogation and torture. Dr. Mitchell smiles and says: "A lot of people just don't have enough faith in what they're able to do."

2. The Power of Self-Healing

For more than thirty years, Dr. Richard Mollica of Harvard Medical School has treated refugees traumatized by war, violence, and torture. His travels have taken him to the darkest places on earth, including the killing fields of Cambodia, the massacres of Bosnia, the genocide of Rwanda, and the ruins of the World Trade Center. Dr. Mollica and his colleagues at the Harvard Program in Refugee Trauma have counseled more than ten thousand survivors of unfathomable brutality. He has helped patients who are physically paralyzed because of their psychological damage. He has worked with women who literally cannot see because of their emotional anguish, a condition called hysterical blindness. In all of his encounters, he insists that he has never met a person without the capacity to overcome suffering. "Never a hopeless patient," he says adamantly. "*Never*. And I don't say this lightly."

In July 1995, the Serbian army massacred eight thousand Bosnian men and boys in a town called Srebenica. Many victims were tortured and then executed with bullets in the head. Mothers were forced to watch their sons slaughtered. Corpses were strewn in the streets. Bulldozers pushed bodies into mass graves, and some were buried alive. Large groups of unarmed men were attacked with machine guns and artillery shells. It was the worst mass murder in Europe since World War II. Dr. Mollica traveled to Srebenica to help some of the women who had lost husbands and sons. The survivors described feeling dead to the world. They didn't want to go on living. There was nothing left for them. And yet, Dr. Mollica says, "I've seen them recover." In his provocative book *Healing Invisible Wounds,* he asks why we are quick to believe that a knife wound to the chest will heal but an injury to the mind will never repair itself. In a biological sense, he believes, there is no difference between an injury to the body and an injury to the mind.

Dr. Mollica calls this "self-healing"—a force inside all human beings "to restore our physical and mental selves to a state of full productivity and quality of life, no matter how severe the initial damage." He is well aware that his views may sound naive or overly optimistic, but he insists that two decades of scientific research support his theories. Above all, he trusts what he has witnessed with his own eyes in the very worst places on earth. "The full flowering of human resilience is awesome," he says. Every human being is born with the strength to heal. No amount of violence or torture can destroy that capacity. No barbarism or savagery can crush the ability to recover and rebuild. "Even in the most hopeless human being," he tells me, "there is hope."

* * *

Part 1 of this book is now complete. We've explored the primary question: What does it take to survive? Clearly, no single theory can encompass who lives and who dies. No common denominator applies to every person or struggle. In some cases, the cosmic coin toss accounts for everything. Alzheimer's patients don't pick their DNA. Trauma victims don't choose the drunk driver swerving down the streets. Still, survival isn't entirely out of your hands. In fact, you control much more of your destiny than you may have imagined. Above all, your mind-set makes the difference. You can take care of yourself and pay attention to your surroundings. You can make your own luck in the worst situations. You can pray, too, if it suits you. And you can persevere with willpower.

After interviews and adventures around the world, I believe there are as many doors into the Survivors Club as there are personalities. What begs the question: Which path will you follow? Do you have what it takes? Will your attitudes and strengths help you survive? Part 2 of this book focuses on *you*. In just a few moments, you'll take the Survivor IQ test on the Internet and instantly receive a custom profile of your crisis personality with detailed answers to the second pivotal question of this book: What kind of survivor are you?

Survival Secrets

The Oldest Living Thing on Earth

Its precise location is a highly guarded secret, but Methuselah lives in the White Mountains of California, a harsh and desolate spot in the shadows of the mighty Sierra Nevada range. At 9,500 feet, it's an extreme environment, rocky and uninviting, where the average high temperature only reaches thirty-six degrees Fahrenheit and the low plummets to twenty-six below. Fierce and relentless winds slap the mountainsides, blowing away most forms of life. The soil is chalky, dry, and barren, and for extra biblical drama, lightning strikes often enough to leave charred wood everywhere. It's a wonder anything can survive here. And yet Methuselah, an ancient bristlecone pine tree, is at least 4,650 years old and going strong.

Stretching more than thirty feet into the air, Methuselah is the oldest living tree on earth and, depending on your definition, may be the oldest living *thing*.* It is named after Noah's grandfather, who lived to the age of 969, the longest lifespan of anyone in the Bible. Some people say the tree looks more dead than alive, but Methuselah is actually in perfect health, enjoys an active sex life, and shows no worrying signs of senility. It sprouted more than three thousand years before the fall of Rome, two thousand years before the Book of Genesis was written down, and six hundred years before Abraham. The tree was already a seedling when the Great Pyramids of Egypt were built, and it has survived volcanic eruptions and glaciers that stretched all the way south to where Los Angeles stands today. Across the ages, it has endured every imaginable assault, including logging and mining during the Gold Rush, atomic testing a hundred miles away in the desert, and waves of tourists who carved out chunks of ancient wood when a 1958 magazine article announced its discovery and location.

Today in the Inyo National Forest, they've taken away all signs pointing directly to Methuselah. They want to

* The bragging rights are hotly contested. In the Mojave Desert of California, an evergreen flowering shrub called a creosote bush is believed to be nearly 12,000 years old; in Tasmania, colonies of King's lomatia, a plant clone, are believed to be over 43,000 years old; and deep below ground in Carlsbad, New Mexico, bacterial spores that are 250 million years old have been found in sea salt crystals.

protect it from vandals, thieves, and tourists who might scratch their initials into its sun-bleached bark. From the forest service visitor center, a two-mile hike leads to the desolate spot where it lives. Somewhere in the distance, over a few mountain ranges, the lights of Las Vegas make the sky glow at night. Methuselah was discovered and named by Edmund Schulman, an early pioneer in the field of dendrochronology, the science of dating historic events like climate change through the analysis of tree-ring growth. Mesmerized by the ancient tree, he carefully counted its rings. When he died of a heart attack at age forty-nine, a memorial service was held in its shadows.*

Schulman revered the bristlecone pines and was amazed by their resilience and tortured beauty. They look like haunted trees from the imagination of Maurice Sendak or Tim Burton. At the base, they're stubby and their withered bark resembles sagging skin, while their jagged branches seem to scratch at the sky as if they are clawing for life. "Longevity Under Adversity" is the title of one of Schulman's scientific papers about the remarkable bristlecones. "The capacity of these trees to live so fantastically long," he wrote, "may, when we come to understand it fully, perhaps serve as a guidepost on the road to [an] understanding of longevity in general." In his final years, struggling with health problems, Schulman even imagined some kind of magic elixir made from these trees that would someday help humans live longer.

Of course, there is no bristlecone panacea, at least not yet. But *longevity under adversity* is a phrase that resonates

---

* Schulman discovered another bristlecone even older than Methuselah but he died before having a chance to study it. The unheralded tree is around 4,950 years old and in excellent health.

with Thomas Swetnam, director of the Laboratory of Tree-Ring Research at the University of Arizona where Schulman worked. Swetnam believes Methuselah's secret is that it grows very slowly—adding just one inch to its girth every hundred years. That makes its wood extremely dense and resistant to the inhospitable environment. Its pine needles offer another advantage. One set can last forty years, and they're specially designed to keep moisture from evaporating. That's particularly helpful, because the tree only gets a drink six weeks a year, during the snowmelt. If the tree's bark breaks, its needles unleash terpenoids, chemicals with a scent that can scare creatures away. It also helps that this brutal patch of earth is hostile to most other threats. It's safe from wildfires because there's so little to burn. It's too tough for insects, fungi, or other enemies. Very few species can live in this place, limiting the competition for resources. Perhaps most astonishing, the bristlecone has the awesome ability in tough times to shut down a limb. "These trees will sacrifice themselves," says Swetnam, explaining that one section will die or stop growing—if water is too scarce, for example—while the rest of the tree stays alive.

How long will Methuselah live? I ask LeRoy Johnson, a retired geneticist with the forest service who worked for eight years with the ancient bristlecone pine trees. The seventy-one-year-old visits the old tree frequently and believes it's "perfectly healthy." Indeed, judging from its sex life, Johnson says, it's as vigorous as ever, producing a few pinecones each year. Johnson took one of them, extracted its seeds, and planted them in a forest service nursery in Placerville, California. Every single seed germinated, an astounding result. There seems to be no stopping this old tree. "Adversity breeds character," Johnson says admiringly. Indeed, he believes Methuselah can teach us all a lesson about survival: "Don't worry, just keep plodding on." Unless the forest service unwisely reveals its location or unless fire or erosion finally bring it down, Johnson believes Methuselah should survive another five thousand years. Theoretically, he says with awe, it could live forever.

PART II

Are You a Survivor?

A survival situation brings out the true, underlying personality. Our survival kit is inside us.

—LAURENCE GONZALEZ

13

The Survivor Profiler

Discovering Your Survivor Personality

Drive two hours north of Spokane, Washington, and you'll end up in the vast wilderness of the Kaniksu National Forest, not too far from the Canadian border. The sky here is brilliant blue and the woods smell like real-life air freshener. When the two-lane blacktop road turns to dirt, you start to see yellow signs nailed to the pine trees:

CAUTION: U.S.A.F. SURVIVAL TRAINING IN PROGRESS

This half-million-acre classroom belongs to the air force. In these mountains where bears, wolves, and cougars roam, the military teaches something called SERE, the abbreviation for Survival, Evasion, Resistance and Escape. In this secretive world, the trainers don't welcome gorgeous days like this one. One instructor tells me they prefer when it's "wet, cold, and miserable." Only in the worst conditions can trainees begin to understand what it's like to be shot down or captured by the enemy. The more wretched and distressful, the better. The air force believes in something called stress inoculation. Training operates like a vaccine: A small challenge to your system is supposed to prepare and defend you against much greater adversity. In other words, if you're exposed to enough hardship and pressure, you'll build your immunity. It's a variation on classic psychological conditioning: The more shocks to your system, the more you can withstand.

So here's what they do in these woods: They deprive you of food and sleep, chase you around like an animal, and when they capture you, they drag you to a deceptively named place called the Resistance Lab. Translation: prisoner of war school. It's strictly off limits to outsiders, and what really goes on behind the concertina wire is classified. Everything inside POW school is modeled on real enemy encampments, including guard towers, barbed wire, concrete cells, and metal cages. To scare you, they've even got fake graves marked with crosses. The goal is to simulate hell on earth like the Hanoi Hilton or al-Qaeda's torture chambers in Iraq. If they even allow a visit to the latrine, you relieve yourself in a putrid hole in the ground. Highly trained professionals serve as jailors and interrogators, putting "prisoners" through carefully choreographed chaos that's designed to disorient and break them down. If you're afraid of dogs, they may terrify you with snarling German shepherds just inches from your face. If you're scared of snakes, they may throw you into a pit of writhing (but nonvenomous) reptiles. If you're claustrophobic, they may stuff you into a series of smaller and smaller boxes or bury you underground in barrels. Throughout the experience, they wear you down with sleep deprivation, semi-starvation, and blaring music, including *Sesame Street* songs around the clock. They interrogate you constantly, employing enemy techniques copied from World War II, Korea, and Vietnam. They don't use excessive force, but the training can be rough. Bones and eardrums have definitely been broken in the service of preparing American men and women for the reality of captivity.

At every stage, the goal is to make you crack, fail in your mission, and then show you how to put yourself back together again. The first phase is called hitting the wall—the moment when you *believe* you can't take another step, when you can't survive another minute, when you're willing to do anything just to stop the misery. The course is carefully designed so that everyone hits the wall. In that critical moment, when you're begging for food and rest, the trainers push you even harder. The purpose is to prove that you're stronger than you know and that you can keep going. Success doesn't depend on your size or strength. Instructors tell me that big, strapping guys often crumble like blue cheese. Women who have gone through the pain of labor and childbirth are often much better survivors than the most muscular and athletic airmen. Survival is all in your mind.

When you see Tom Lutyens strolling through the forest, you would never know he's a legendary instructor in the world of survival. Indeed, he's so valuable—his expertise unpar-alleled—that when he retired in 1988, the air force kept him working as a contractor. Tom is sixty-one years old and six and a half feet tall. He wears olive wool trousers, black hiking gaiters up to his knees, and a blue-and-green flannel shirt made by his mother. With a high forehead, hair swept back, and a fascination with wildlife, he reminds me of Rex Harrison's Dr. Dolittle, the character who talks to the animals. He kneels next to a huge pile of fresh scat and inspects it gleefully. It's a clear sign the bears are coming out of hibernation. When we hear a drumming sound in the woods, Tom proclaims, "There's dinner!" It's a grouse—relatively easy to catch, he says, and delicious roasted over a campfire. Soon we see a procession of red ants crossing the trail. "They're delicious," he cries pinching one between his long fingers. "They taste lemony."

Tom fires off a barrage of survival advice. *Practical:* Peanut butter is "the best survival food" in terms of cost per calorie. *Impractical:* A rabbit's eyeballs are a good place to find salt for survival. *Practical:* Always carry a mini flashlight, because you never know when the electricity will go out. *Impractical:* Always carry a metal match—technically, a ferrocerium rod—in order to start fires even in the rain. *Practical:* Bandannas are versatile accessories because they can protect you from the sun, stop bleeding as a tourniquet, filter dirty water, trap fish or animals, and even kill as a garrote or sap. *Impractical:* Blue jeans are dangerous in cold, wet weather because "cotton kills" by trapping moisture and accelerating hypothermia.

The suggestions keep coming, but above all Tom wants you to know that survival isn't only for air force men and women who have trained in these woods. Survival is an outlook that anyone can learn, practice, and apply. At some point, everyone experiences a "dislocation of expectation," the air force term for shock. Who lives and who dies? Tom believes that 80 percent of your survival depends on your attitude; 10 percent on what you know and how you apply it; and 10 percent on the tools and equipment you've got to deal with your challenge. In other words, you can learn all sorts of skills and carry all kinds of equipment, but in the end your survivor personality will play the deciding role. "The best survival kit is between your ears," he says.

1. The Survivor Profiler

What if you could discover your survivor personality *before* the cement truck barreled around the corner? What if you could uncover your top survival tools *prior to* needing them? I'm not talking about testing your knowledge of fire starting, shelter building, or celestial navigation. I mean something else: What if you could identify your main psychological strengths *in advance* of your next crisis? As I began to ask these questions, I quickly found myself immersed in the fascinating field of psychometrics, the science of psychological measurement. I discovered that modern personality testing is extremely sophisticated, and the results are very revealing. I also met a woman in Lincoln, Nebraska, named Dr. Courtney McCashland, the founder and CEO of TalentMine. McCashland helps some of the world's biggest companies identify and develop the strengths of their employees. She honed many of her skills working for the legendary Gallup organization, where she developed personality tests that have been taken by millions of people. Her specialty is discovering what you're really good at doing and teaching you to nurture and build those strengths so that you can become more successful, productive, and, ultimately, happy. To measure your talents, McCashland designs custom Internet-based surveys and interviews that capture your spontaneous thoughts, feelings, and behaviors. Next, using complex equations called algorithms, she sorts through all of your answers, finds the key patterns and traits of your personality, and generates a powerful and revealing profile of your strengths. After taking her tests and hearing her insights, you can't help asking: *How did she know that?*

McCashland isn't just a psychometrician. She's also a co-survivor. In 1987, she lost her beloved big brother Curt in what appeared to be an intentional drug overdose. In 2001, her one-year-old son Noah was diagnosed with leukemia. After three years of intense treatment and some scary times, Noah is happy, healthy, and cancer-free. When I approached her about my book, she had never designed a test to measure the survivor personality. She searched far and wide to see if anyone had ever tried, but she couldn't find anything specifically aimed at survivorship. There are many excellent tests—known as instruments—that measure resilience, hardiness, and coping skills, but none examines the many different strengths necessary for survival. So I partnered with McCashland and TalentMine to create a brand-new test called the Survivor Profiler. The result is an easy-to-use fifteen-minute questionnaire that will reveal what you've got inside to handle life's toughest challenges.

You may find it hard to believe that an Internet test can produce a meaningful portrait of a person's nuanced, sometimes contradictory psychology, but that's exactly what the Survivor Profiler can do. It's based on McCashland's years of experience studying how successful people get things done in life and work. The one thing these individuals share, McCashland says, is an awareness of their strengths. High performers make a point of channeling time and energy around their talents. By leveraging these abilities, they accomplish more, feel healthier, and are more engaged and satisfied in life. McCashland argues passionately that every one of us can identify our survivor strengths, exercise and build them like muscles, and mobilize them in a crunch. That's the entire point of the test you're about to take.

2. How the Survivor Profiler Works

Modern personality tests are based on real-world interviews with people about their lives and work. They're powered by advanced computers and formulas. They're designed using state-of-the-art psychology and statistics. They're engineered so that it's very hard to cheat and beat the system. The level of accuracy and reliability is usually very high. And the results are often quite surprising. To be clear, these tests aren't like those quizzes in magazines and newspapers. Psychometric tests like the Survivor Profiler are powerful and sophisticated tools to help you know yourself better.

To develop the Profiler, McCashland and her team searched for recurring patterns and themes in the experiences of survivors I interviewed. While the details of their ordeals varied greatly, it was clear that the most effective survivors almost always draw on the same set of psychological tools. Obviously, everyone is different and each struggle is unique, but successful survivors can be divided into discrete groups. All told, McCashland and I identified five main kinds of survivors and twelve essential tools that they use in the worst situations. During the development phase of the test, McCashland found that virtually everyone fits into one of the five survivor archetypes, and each of us possesses at least three of the twelve key tools.

In the personality test you're about to take, you'll be asked to consider a series of paired statements like the example below:

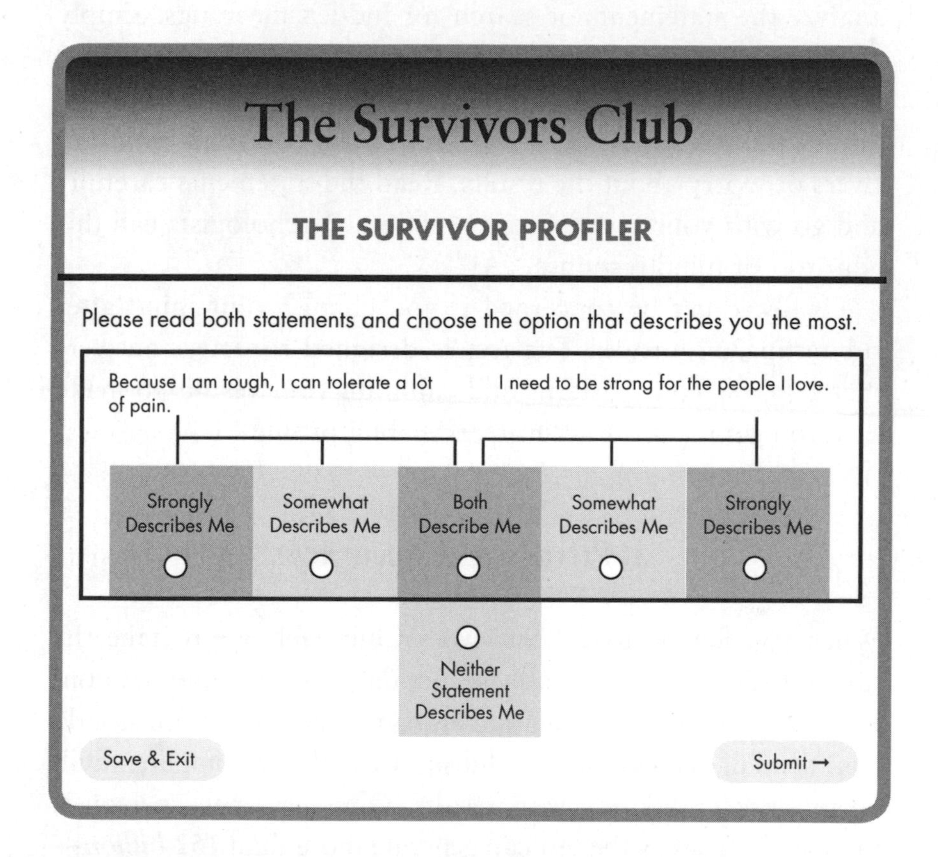

After reading each statement, you're supposed to select the option that best describes you. In the example above, are you tough and able to tolerate a lot of pain? Or do you need to be strong for the people you love? Sometimes it may be difficult to select between the two statements, because they both capture you well. Don't worry—that's the way it's supposed to be. In fact, McCashland deliberately paired statements that aren't exact opposites. If they capture you equally well, then check the circle in the middle: *Both Describe Me.* In other pairings, you may feel no connection to the statements at all. That's why you're given the bottom choice: *Neither Statement Describes Me.*

It's important to emphasize that there is *no* right or best answer to each pairing. As you go through the test, try not to overanalyze the statements or search for hidden meanings. Simply pick the option that *feels* the best to you. In a survival situation, you may not have a lot of time to evaluate your choices. Approach the test in the same way. Try not to overthink your answers or worry about the results. Read the statements carefully and go with your spontaneous reaction. Psychologists call this your top-of-mind response.

Finally, don't be concerned if you change your mind after submitting an answer. The test is designed to detect answers that don't fit your patterns and calibrate your results to make sure you get the most accurate, revealing profile.

3. Privacy and Accuracy

When you log on to the Survivors Club Web site to take the test, you'll connect to a highly secure data center in Boca Raton, Florida. All of your information and answers will remain strictly confidential. As soon as you finish, a cluster of computers will go to work calculating your results. Obviously, each person is unique. That's why the test can generate more than 152 *billion*—that's 152,000,000,000—different profiles. The computers will tally your answers, find your patterns and strengths, and within seconds produce a customized report on your survivor personality. Your profile will include several parts: (1) Your Survivor IQ, which tells you what type of survivor you are; (2) your top three Survivor Tools; and (3) several in-depth profiles of real-life survivors who closely match your survivor personality. Once you receive your Survivor Profiler report, you'll have a few options: You can read it on your screen; you can save it in your computer; you can e-mail it to yourself; or you can print it out.

Accuracy and reliability are very important. McCashland asked a number of leading psychometric experts to review the test and its underlying thinking. The development process showed the test to be highly dependable and accurate. The industry standard for reliability on these kinds of tests is 0.7 on a scale of 1.0. The Survivor Profiler's score was an impressive 0.92.

4. Will I Fail?

Full disclosure: I've never been a very good standardized-test taker. To be even more frank, I've often worried about flunking. So let me reassure you: It's *impossible* to fail this test. There are no grades. You won't receive a good or bad score. As you know by now, the first rule of this book is that *everyone* is a survivor. The corollary is that *everyone* comes equipped with specific tools to handle adversity. The Survivor Profiler will identify them for you by measuring twelve different dimensions of your personality. In psychometric terms, it's looking for the frequency and intensity of twelve survivor strengths. It's true that some people possess more of these strengths than others, but when you take this test, you'll discover that you've got plenty of tools in your kit.

5. Will the Results Make a Difference?

In chapter 8, you briefly met Dr. Al Siebert as he described the serendipity personality. Siebert has devoted his entire career to investigating survivorship. In 1965, after finishing graduate school at the University of Michigan, he began to analyze "the Survivor Type Person." He's in his seventies now and lives on the Columbia River near Portland International Airport. It's a convenient location because he travels the world lecturing on "the resiliency advantage" and "highly resilient survivors." Siebert tells me that survivors are ordinary people like you and me with flaws, worries, and imperfections. Only a few of us are natural-born survivors. "Just as some people are born musicians or artists, some people have a natural talent for coping well," he writes in *The Survivor Personality.* "The rest of us need to work consciously to develop our abilities." The most effective survivors, Siebert believes, are people who are able to learn from their knocks in life. When confronted by crisis, they tap into their deepest strengths and abilities. "Fortunately," he writes, "almost every person is born with the ability to learn how to handle unfair situations and distressing experiences."

The Survivor Profiler is designed to help you identify your strengths. It's based on the experience of successful survivors around the world. It identifies and measures tools that have enabled ordinary people to overcome extraordinary adversity. I would never be so naive as to suggest that a personality test can save your life. But I'm confident the Profiler can help you channel your talents and give you an advantage in a crisis. Experts agree that planning and preparation are the twin pillars of survival. If you bring a compass and map—or GPS—when you go hiking, you've got a potential survival edge over someone who packs only a picnic. The same applies to your inner resources. The Profiler gives you the manual for your Survivor Personality and the tools to navigate your next challenge.

6. How to Get Started

If you slip off the dust jacket of this book, you'll find your unique access code printed on the reverse side. It has ten characters and looks like this:

E8U3JR4BU8

Please make a note of your access code and then go to the following Web site:

www.TheSurvivorsClub.org

You'll find a clearly marked Survivor Profiler section and welcome page. The instructions are easy to follow. Your user name will be your e-mail address. To ensure the confidentiality of your answers and results, you'll be asked to create a password. Please make a note of it. You can take the test on virtually any computer with access to the Internet. Sometimes certain computer settings like pop-up blockers will prevent access to the test. You'll find instructions on the Web site about how to change those settings safely so that you can take the test. If you have any problems, you can e-mail for technical support. As I mentioned, most people complete the test in around fifteen minutes. Don't worry if you're interrupted. You can stop at any time and resume where you left off.

The next two chapters will serve as a kind of manual for your Survivor Profiler report. I'll begin by explaining the five types of survivors. Then I'll tell you about the twelve key survival tools. If you're ready to get started, just find your access code on the dust jacket and log on to www.TheSurvivorsClub.org.

14

Your Survivor IQ

What Type of Survivor Are You?

It's not easy to categorize people. After all, there are 6.6 billion of us on earth, and everyone is unique. But after studying the personalities and patterns of people who overcome adversity, five main Survivor Types emerge. They are:

The Fighter
The Believer
The Connector
The Thinker
The Realist

What category describes you best? If you've taken the test, then you know your Survivor Type. You've also received an in-depth portrait of a real-life survivor who exemplifies your category. In the pages that follow, I describe each kind of survivor. You may be wondering if you've got the *best* Survivor Type. That's natural. Let me allay your concerns. Sure, some people might be better suited to certain survival challenges, and you might benefit from certain strengths in specific survival situations. For instance, Fighters may be better equipped to survive a hostage situation or violent crime. But Believers are effective, too, and many ex–prisoners of war like Senator John McCain told me that faith was their most important weapon in captivity. And it's critical to emphasize that one type isn't any better than another. Each has different strengths, but *all* have what it takes to survive. It's also essential to point out that some people may fit into more than one group. In fact, you may recognize different aspects of your personality in each of the Survivor Types. The Profiler takes this into account when it calculates your Survivor IQ.

You may also wonder if it's possible to change your Survivor Type. The answer is probably not. If you take the test more than once, your results probably won't differ much. In fact, during the development phase, we found very little variation among repeat test takers. In terms of your personality, psychologists say, it's not easy to change who you are at the deepest levels where spontaneous thoughts, actions, and beliefs originate. That said, it's definitely possible to make changes in your attitudes and behaviors and to develop new abilities to cope with adversity. If you're interested in survivorship information and professional resources, please visit www.TheSurvivorsClub.org. As a journalist and writer, I'm hardly in a position to offer psychological advice. I leave that to the experts.

What follows are descriptions of the five Survivor Types.

THE FIGHTER

Fighters come in every shape and size. To be a Fighter, you don't have to punch like Muhammad Ali or refuse to surrender like John McCain. Fighters attack adversity head-on with purpose and determination. Against any odds, they're driven to succeed and won't stop till they achieve their goals. When you're a Fighter, you never stop attacking. Even at your lowest, you still find a way to bounce back and counterpunch. You have a passion for life and seize every day with zest and zeal. You've got the willpower and determination to struggle, resist, and overcome even in the face of formidable opposition. Maybe you're courageous and brave. Maybe you're aggressive and competitive. Maybe you're stubborn and unyielding. No matter, you get pumped up by the heat of battle. You push yourself to be the best. You're motivated by a sense of purpose or a calling greater than yourself. You're here on earth for a reason that's worth real sacrifice. You're resilient, you're tenacious, and you often feel stronger because you've endured hardship in the past. When you get knocked down, you bounce back again. You're indomitable, psychologically tough, and you can endure more physical pain and suffering than most. You keep going when others have given up and you battle to the very end. Above all, you're a Fighter.

THE BELIEVER

When you're a Believer, you put your faith in God to protect and sustain you through your trials. Your beliefs and convictions are like life preservers keeping you above water in difficult times. You trust deeply that God has a plan for your life and will steer you through any adversity. You're convinced the Lord would never give you a challenge you couldn't handle. Your upbeat spirit lifts you when others are down. Even in the worst times, you feel blessed and are confident things always work out for the best. Even if death approaches, you find comfort knowing that God loves you and your life is in His hands.

You draw remarkable emotional and physical power from your faith. Strengthened and guided by God, you feel capable of shouldering almost any burden. If you call out to Him and open yourself to His wisdom, you know He will answer. This faith also gives you optimism and hope, powerful weapons in survival. You're able to banish negative feelings or flip them around into positive thoughts. You can find humor in the darkest times and even laugh in the face of adversity. You are strengthened and emboldened by your faith and comforted by the conviction that your fortunes will improve one way or another on this earth or in the next life. Above all, you're a Believer.

THE CONNECTOR

When you're a Connector, you overcome incredible adversity with the power of your relationships and bonds with other people. You are deeply devoted to your family and friends. Your love for your parents, spouse or partner, children, and friends motivates you to tackle enormous obstacles. You know that your family and friends depend on you and need you. You hold these relationships sacred, and you will go to any lengths to protect and preserve them. You draw strength from these primary relationships and often rely on support groups or social networks to help you through difficult times. You're able to lean on others for aid, and you know how to reach beyond your regular circle of friends to find the help you need. You're a good networker who makes the most of your connections. You often feel great empathy for others who are struggling. You take care of other people before you look after yourself. You're good at reading strangers and situations. You know how to get along with others. You play well on teams and work effectively with others to get things done. You survive because of your powerful bonds. You would endure anything—and do everything—for the people you love. Above all, you're a Connector.

THE THINKER

When you're a Thinker, you use your brain to overcome your obstacles. Your intelligence has many dimensions. You rely on a combination of smarts, creativity, and ingenuity to solve problems. Book learning isn't your only resource. You've also got street smarts and common sense. You see your challenges clearly and are good at diagnosing the underlying nature of a problem. In tough times, you look at all the angles, generate new ideas, and discover unexpected solutions. You don't get distracted easily. You're highly focused, analytical, and rigorous; you concentrate on what needs to get done. Your mind is practical, not up in the clouds. You're good at turning ideas into action. When others get stuck, you can improvise and find a way out. Logic and reason help you understand the real facts of your situation and the consequences of your choices. Common sense helps you apply your knowledge and experience in creative and productive ways. While some people depend on muscle and brawn to win life's battles, you rely on your mind. Above all, you're a Thinker.

THE REALIST

When you're a Realist, you recognize that everything doesn't always go as planned. You take life as it comes, knowing that you can control some things but not everything. Instead of resisting or fighting, you make the most of your situation and go with the flow. When others overreact or panic, you stay calm and collected. When faced with a challenge, you're pragmatic, quickly figuring out the best way to cope. Your reactions are instinctive and practical—they occur almost without thinking—and you have immediate insight into your problems. Intuitively, you know how to sit back and wait for the worst to pass. You also know when the moment is right to take action. You're confident in your ability to do what you need to do. In a crisis, some people lose sight of reality. They get overwhelmed. They're overly pessimistic or optimistic. But you're the opposite. You deal with facts and what's really in your control. You survive by riding out the storms of life and doing what's necessary to keep going. Above all, you're a Realist.

15

Your Survival Tool Kit

What Are Your Top Three Strengths?

Your Survivor Profile report is divided into two sections. The first part explains your Survivor Type. It paints the big picture of your survivor personality. The second part digs deeper into your psychology and tells you your top three Survivor Tools. The most effective survivors draw upon a common set of psychological strengths. What follows are the twelve that are measured by the Profiler:

Adaptability
Resilience
Faith
Hope
Purpose
Tenacity
Love
Empathy
Intelligence
Ingenuity
Flow
Instinct

While no list can capture *every* survival tool that can help in a crisis, these twelve showed up most frequently in my interviews with survivors and discussions with experts. They also made the greatest impact in the most difficult situations. The Profiler calculates your top three tools from these twelve. You probably have many other abilities at your disposal. The three that are described in your report—the top 25 percent—are your most dominant strengths, but they're not necessarily your only ones.

When you study your profile, you may wonder whether the order of your tools matters. The answer is yes and no. Yes, because the Profiler ranks your strengths in order of their prevalence. No, because the statistical differences in the rankings may be very small. The formulas that generate your survivor type and tools are designed to check and cross-check the patterns to make sure you're receiving a meaningful profile of your personality. So the order matters in a technical sense but not in any fundamental way.

Over and over, experts told me it's not *what* you know that makes a difference in survival. It's *how* you apply it. Pure knowledge—the kind from books and lectures—only gives you the *possibility* of action. It's potential energy. It's incomplete until it's put to real use. Applied knowledge is the key to survival. It's kinetic energy. It turns action into *reality.* It often makes the difference between living and dying, winning and losing.

ADAPTABILITY

Your adaptability is a critical survival tool. You've got the capacity to adjust readily to different situations and to change your attitude and behavior to handle new challenges. In a crisis, you're flexible. When you confront an obstacle, you're able to modify your approach to accomplish your goals. You quickly let go of the way things used to be and tailor your strategy and tactics to fit the new reality. It may sound simple or easy, but not everyone can do this. Indeed, when adversity strikes, many people refuse to accept their new circumstances. They go into denial. They act like nothing has changed. They pretend everything is okay. They're rigid and unbending. You're the opposite. You're attuned to your dynamic surroundings. You recognize that life brings ups and downs. It may not be your preference, but you make necessary changes in order to prevail. You find new ways to cope with your problems.

For sixty thousand years, our success as a species has depended on adaptability. Indeed, the famous phrase *survival of the fittest* wasn't intended to connote physical health or vigor. It was coined by Herbert Spencer in 1864 to describe the suitability of a species in an environment. Charles Darwin borrowed the concept in *The Origin of Species,* writing: "In the struggle for survival, the fittest win out at the expense of their rivals because they succeed in adapting themselves best to their environment." In everyday survival, that translates into the ability to adjust to adversity and fit into new and challenging situations. More than any other psychological strength, adaptability determines who gets into the Survivors Club.

RESILIENCE

An African proverb says the wind does not break a tree that bends. This wisdom describes you: You have the ability to bend but not break. The word *resilience* comes from the Latin *resiliens*—to rebound or recoil—and it it's defined as the power to return to an original form after being twisted, compressed, or stretched. In science, it means elasticity, the capacity of a substance to change its shape in response to a force and then recover. In survival, it means you can bounce back when you're knocked down. In difficult times, you pick yourself up, dust yourself off, and persevere. You're accustomed to hardship. You've overcome rough times and you're stronger and smarter because of what you've endured. When others give up, you keep trying. When others get tired, you push forward and refuse to quit.

Around the world, people have always seized on symbols or talismans of resilience. In Japan, one of the most popular good-luck charms is the Daruma doll, a red-and-white papier-mâché creation that represents the Buddhist monk Bodhidharma, founder of Zen, who meditated in a cave for so long that he lost the use of his arms and legs. Daruma dolls have no limbs and they're weighted at the bottom so that when they roll over, they pop back up. They're collected as good-luck charms and symbols of perseverance. The Japanese say *nana korobi yaoki:* Fall down seven times, stand up eight. In a phrase, that's you.

FAITH

Faith is the most powerful and universal survival tool. Your faith means you trust that God has a plan and will look after you. You believe that a greater power will steer you through difficult times and guide your actions. If you listen, God will show you the way. In tough times, you're less afraid because God loves you and will always take care of you. In a crisis, faith gives you remarkable power and confidence to prevail.

You don't worship a distant, aloof God. You believe in a higher power who is loving, caring, accessible, and available. "This is a God who listens to prayer, who responds, who desires good for humanity," Dr. Harold Koenig writes in *The Healing Power of Faith*. You trust God to "fill the gap" between the challenges you face and what you're capable of handling, Dr. Koenig explains. You are never alone in your struggles. You see God as your "active partner" in every challenge you encounter. Faith is your greatest comfort and mightiest weapon.

HOPE

Hope gets you through the toughest times. You believe that no matter how bad life gets, everything will turn out for the best in the end. When you have a wish or desire, you are confident that it will be fulfilled. Your kind of hope isn't a rosy sunrise on a Hallmark card. It's a combination of optimism and realism. You believe good things happen, but you aren't naive.

Your hope lifts others when they're down. You're an upbeat person who can turn negative feelings into positive thoughts. You feel lucky and blessed in your life. You believe hard times never last too long and always make a turn for the better. You can also laugh in the toughest times and see the humor in even the darkest situations.

Dr. Jerome Groopman, a best-selling author and Harvard Medical School professor, has carefully investigated the science of hope. He found that your brain pumps chemicals responsible for the hopeful sensation, which in turn block out pain and accelerate healing. Belief and expectation—the one-two punches of hope—release neurochemicals called endorphins and enkephalins that mimic the effects of morphine. As a result, "Hope helps us overcome hurdles that we otherwise could not scale," Dr. Groopman writes, "and it moves us forward to a place where healing can occur." Hope doesn't directly lead to recovery, he notes. But it contributes to survival. In scientific terms, they may not be causally connected, but they are definitely correlated. Whenever Dr. Groopman meets new patients, reviews their medical history, and performs a physical exam, he writes in *The Anatomy of Hope,* "I am doing more than gathering and analyzing clinical data. I am searching for hope. Hope, I have come to believe, is as vital to our lives as the very oxygen that we breathe.

"I see hope as the very heart of healing," Dr. Groopman continues. "For those who have hope, it may help some live longer, and it will help all to live better."

PURPOSE

Purpose is the booster rocket of survival. It gives you the power and drive to persevere in the face of incredible adversity. You have big goals that you strive for—they're the reason you're alive, and they make every day worthwhile. You have a passion for life and your dreams. You believe that life is a gift and it's up to you to make the most of it. You're determined in the face of adversity and focused on accomplishing your objectives. You confront your challenges with unwavering conviction. You have a mission in life, a reason for living that is greater than yourself. You are driven by a profound sense of duty to a cause, whether it be God, country, family, or friends. You work tirelessly and are willing to sacrifice deeply for your purpose and principles.

Viktor Frankl, the eminent psychiatrist and Holocaust survivor, observed that "everything can be taken from a man but one thing: the last of the human freedoms—to choose one's attitude in any given set of circumstances, to choose one's own way." Quoting the nineteenth-century German philosopher Friedrich Nietzsche, Frankl added: "He who has a *why* to live for can bear with almost any *how*." Purpose gives you the why—the meaning and mission—in your life. It also gives you the power to survive.

TENACITY

Tenacity comes from the Latin *tenacitas,* the act of holding fast. Your tenacity is your superglue in the toughest times. You've got the persistence and determination to stick to it when others let go. You've got the grit and toughness to hold on when others can't take the punishment. Sure, life hurts everyone, but you can absorb pain and suffering and keep going. You have an indomitable spirit and can endure real hardship. When others give up, you push yourself even harder. In a crisis, you're willing to go beyond any limits. You always rise to the challenge and take pride in your ability to withstand the worst. That's you: You never let go.

LOVE

The capacity for love is universal, but your level of devotion is different. You will do anything—and go to any lengths—for the people you love. Your bonds with family and friends are unbreakable and give you the reason for living. Purpose—a survivor tool described earlier—is defined as a calling greater than yourself. For you, love is the ultimate purpose. In a crisis, you always think about your family and friends first. Other people depend on you, and you can never let them down. Your attachments make your life worthwhile. In tough times, you're less afraid because you're surrounded by the love and support of family and friends. In an emergency, you would walk through fire for the people who depend on you. Even in the worst times, the love around you and inside you gives you the strength to go on.

EMPATHY

Empathy is a surprising survival tool. It may seem counterintuitive, but in a crisis, your ability to help others turns out to be a very powerful way to help yourself. Your compassion motivates you to help other people stricken by misfortune. You feel a deep connection with others and will do anything to reduce their suffering. You take care of others before you look after yourself. You're sensitive to other people's needs and go out of your way to look after people you don't know. You're good at reading new situations and people, and you're always aware of your surroundings. In group situations, you work well with others. You're a team player. In a crisis, some people resort to selfishness to survive. But you're different. Your empathy and altruism are far more powerful. By caring for others, you take care of yourself.

INTELLIGENCE

You've got brains and a talent for learning, thinking, understanding, and problem solving. You may be book-smart with a perfect grade-point average, or your intelligence may be more practical and down to earth. As humans, we survived and dominated the planet because of our brains. Our ability to acquire and apply knowledge made all the difference in defeating the saber-toothed tiger and countless other dangers. Your intelligence is a powerful weapon against everyday adversity, too. You see complex and dangerous situations very clearly. You examine problems from all angles in search of realistic solutions. In a crisis, you're good at figuring things out and knowing what's going to happen next. When others get stuck or confused, you're able to break problems into manageable parts and find good solutions. In the worst times, you know that IQ isn't enough. You need common sense and street smarts. You understand people and situations. You diagnose problems, figure out what you need to do, and get things done.

INGENUITY

You are clever, inventive, and resourceful. You know how to apply things you've learned elsewhere to overcome immediate challenges. When others freeze or get stuck in a rut, you see new ways to solve problems. You're good at improvising and finding answers. You enjoy innovation and invention, and you find novel uses and applications for everyday objects. In a crunch, you're a real-life version of Angus MacGyver, the 1980s secret agent on television. In the first episode, MacGyver foiled a missile with a paper clip and created explosives with a common cold capsule. He also stopped a sulfuric acid leak with milk chocolate and thwarted a laser beam with parts from a smashed pair of binoculars.

The proper term for MacGyver's gift is *bricolage:* the art of building things from whatever materials are available. From Mozambique to Taiwan and across the world, the word *MacGyver* is now synonymous with the Swiss Army knife and duct tape, his two favorite tools for improvisation. You may not be a secret agent, but you've got the gift of improvisation and ingenuity.

FLOW

When moving water encounters a rock, it can go over or around the obstacle. Eventually, the river will smooth the stone and ultimately wear it down to nothing. The same applies to you. You move forward, steadily, relentlessly, and with apparent ease and effortlessness. Adaptable people quickly change their attitudes and actions to fit their new circumstances. Resilient people bounce back rapidly from adversity. But you're different. You've got flow. You don't need to make adjustments: You sail along, freely and calmly without fuss or muss. You don't need to pick yourself up when you're knocked down: You stay down, take adversity as it comes, and remain confident that you'll eventually get where you're going. You stay cool when others panic. You relax when others stress out. "As history shows," Laurence Gonzalez writes in *Deep Survival,* "the harder we try, the more complex our plan for reducing friction, the worse things get." You understand the futility—and danger—of trying to control the uncontrollable. Facing a crisis, some fight and others flee, but you flow.

INSTINCT

You have a remarkable gift that isn't learned or taught. You have the innate power of instinct and intuition. You don't need to think very hard—you don't panic or obsess—you simply act. Your gut feelings come naturally and automatically. You trust yourself to do what's necessary. Often you don't know why or how you make a particular decision. It just feels right. Sometimes you notice warning signals that other people miss. Again, it's not logical or intellectual. It's instinctive. In a crisis, you gain immediate insight into your challenge and know what to do. You sense when things aren't right and act on your hunches. In uncertain situations, you trust your instincts. You're highly attuned to everything around you. You see signs of danger before anyone else.

In his best-selling book *The Gift of Fear,* Gavin de Becker observes that the root of the word *intuition* comes from the Latin *tuere:* to guard, to protect. De Becker is one of the world's top experts on threat assessment and runs a successful firm that helps companies and individuals predict and manage violence. The cornerstone of safety and survival, he believes, is your intuition. "You too are an expert at predicting violent behavior," he writes. "Like every creature, you can know when you are in the presence of danger. You have the gift of a brilliant internal guardian that stands ready to warn you of hazards and guide you through risky situations."

AFTERWORD

How to Eat an Elephant

The Lessons of the Survivors Club

The navy taught me to escape a sinking helicopter. The air force trained me to make a deadly weapon with a bandanna. The FAA showed me how to run and jump from a smoke-filled airplane. I mastered the magic numbers of survival and memorized the Theory of 10–80–10. Fifteen feet underwater—or in any crisis—I understood the importance of maintaining a point of reference and waiting for the violent motion to stop. I was nearly finished writing this book when all of that know-how was really put to the test. At two o'clock one morning, I was sound asleep in my home when the burglar alarm went off. "Kitchen door!" an urgent voice intoned from the security control panel. "Alarm!"

My wife, Karen, was lying next to me. My son Will was nestled in his crib across the hall. This wasn't a drill. This was really happening. And I did absolutely everything wrong. First, I froze. More precisely, I sat up in bed and listened for sounds of an intruder. In the literature of disasters, this is known as milling. The dictionary says that means "to move around in churning confusion." You've seen it happen. When a fire alarm blares at the office, people poke their heads into the hallways. Is there really a danger? Is it just a false alarm? Should we evacuate or go back to work? With our burglar alarm going off in the middle of the night, I'd like to think I did what I'm trained to do. I resorted to journalism and began asking questions. Who is in my home? What are they doing? Unfortunately, I was so consumed by news gathering that I tuned out my wife. She was trying to tell me something important, but I was unresponsive. You might say I was a statue in the storm experiencing "behavioral inaction" or "negative panic."

When I finally heard my wife, she was adamantly insisting that we turn the alarm off. Her maternal instinct had taken over. She wasn't especially worried about a thief. She was more concerned about our son. She didn't want the alarm to wake or frighten him. I argued that we should let the siren scream in case there really was an invader in the house. Eventually, the security company would get the signal and send a patrol car to help. After quite a long stretch of doing nothing, I relented and turned the system off.

Next, I decided to call the alarm company—except I didn't have the number handy. So I dialed information and when I finally reached the call center, I demanded to know why our alarm had not triggered any response. I learned to my dismay that the security company didn't have accurate contact information for our home. "Send a patrol car right away," I insisted, trying to sound totally in control. That's when they put me on hold.

It was time to dial 911.

Eventually, the alarm company sent a patrolman with a flashlight and gun, who quickly determined that the kitchen door was ajar. It had been left unlocked, and the wind probably nudged it open. Later, the police arrived. Safe, sound, and mortified, I stood in our front yard and seized the moment. "What should I do if the alarm goes off in the middle of the night?" I asked. First and foremost, the policeman responded, barricade yourself with your loved ones in a safe place. Next: Call 911 and wait for assistance. Even if you're brave, you should resist the impulse to go looking for the cause of the alarm. You may have left the kitchen door unlocked. Or you may encounter an armed intruder on drugs. Bottom line: It's safer behind a locked door in your bedroom, bathroom, or closet. Finally, don't turn off the alarm. The noise can scare prowlers away, and many security systems are connected to call centers that can send help.

Back in bed under the covers, my wife and I marveled at the experience. I was writing a book that asks: Who gets into the Survivors Club? Now we knew the answer: *Not us*. We had never discussed what we would do if the alarm went off in the middle of the night. We weren't prepared with emergency contact information. The security company didn't know how to reach us. And we didn't have a clue about barricading ourselves.

I learned a lot working on this book, and one lesson is this: You never know how you'll respond in a crisis, but mindful-ness gives you a big head start. Sometimes it can take a false alarm to wake you up to this reality. As we've seen, many survivors share a mind-set. In a crisis, they're alert, engaged, and aware. They think—they plan—and they take action. They're not crazy or obsessive. They don't lug survival kits everywhere, although it never hurts to keep one in your home and car. They just pay more attention to staying alive. If you don't think you've got this attitude, it's not too late. The survival mentality is both a reflex and a habit that you can cultivate through practice and preparation. When survivors go to the movies or ball games, they notice the exit signs. When they stay in high-rise hotels, they make a note of the fire escapes. When they face tough medical problems, they educate themselves with facts and different opinions. Psychologists call them information seekers. They learn from life and from their mistakes. If they screw up—like freezing in bed when the alarm sounds—they take advantage of the experience for the next time.

I sleep a little easier now. We've got our emergency contact information nearby. We're all set with the security company. We're ready to barricade and wait for help. And we've even got two new baseball bats at the ready just in case. One is for me and one is for my wife. In a pinch, I'm counting on her. She's got a great swing.

In these final pages. I offer a few last lessons from the Survivors Club. They focus on your mind-set when the alarms in life inevitably go off. They're based on my interviews with people who have overcome the worst. These are their ideas for handling ordinary *and* extraordinary challenges. This wisdom worked for them, and somewhere down the road, they insist, it will work for you, too.

1. One Bite at a Time

At least 80 percent of us respond the same way in a crisis: We're simply overwhelmed. The challenge seems too great, the obstacle insurmountable, the problem unsolvable. In air force survival school, they teach you to conquer this confusion with an axiom that is both colorful and memorable: *You eat an elephant one bite at a time*. Survival is one big ornery animal, they explain, and if you try to devour a fifteen-thousand-pound pachyderm all at once, two things can happen. Either you'll give up or you'll get really bad indigestion. The key to survival is to slow down. Take one small bite. Chew. Swallow. Then take another.

In almost every interview, survivors reinforced this lesson. They used different language and metaphors, but they all spoke of dividing unwieldy challenges into achievable tasks. Nando Parrado, the Andes plane crash survivor whom you met in chapter 1, remembers attacking each crisis by chiseling it into chunks. One goal at a time. One decision at a time. One action at a time. When a devastating avalanche buried the survivors in the airplane fuselage, they whispered to one another: "Breathe. Breathe again. With every breath you are alive." It was literally all they could do. In a hopeless situation without oxygen or light, this approach kept them going until they found a way out.

2. The Illusion of Isolation

My grandmother Lea Romonek was a great lady and a survivor. She was tiny—less than five feet tall—but a real powerhouse. She graduated Phi Beta Kappa from the University of Nebraska and raised two children. She lost her husband but she dated and danced into her nineties. She lived to ninety-eight and a half. Buggy, as we called her, was both practical and wise, the proud descendant of talmudic scholars from Eastern Europe, and when bad things happened or her friends passed away, she always remarked: "We hang by a thread."

For most of us, the thread is only an abstraction. For survivors, it's real. Those who have literally clung to life with their fingernails see each day differently than the rest of us. Alison Wright is a gifted documentary photographer and writer from San Francisco. In January 2000 while traveling in Laos, the tiny Southeast Asian nation, Wright was riding on a bus when it was smashed to smithereens by a logging truck. The crash broke her back, pelvis, coccyx, and ribs. Her arm pierced a window; her spleen was sliced in half; and her heart, stomach, and intestines ripped loose and ended up lodged in her shoulder. She nearly bled to death on the scene. A local teenager sewed up her arm without painkillers, and a passerby drove her eight hours over a bumpy road to a hospital in neighboring Thailand.

Alison is a remarkably strong and independent woman who is happiest gallivanting around the globe, taking photos of monks who rescue tigers in Thailand, gauchos on horseback in Argentina, and penguins on icebergs in Antarctica. Since age sixteen when she left home in Watchung, New Jersey, she has always fended for herself. "Everything in my life was leading up to that moment," she says of her nightmare in Laos. "You stockpile that stuff." Above all, her near-death experience and countless scrapes around the world have given her an affinity with people who are suffering. "The wonderful thing about being a member of this [survivors] club," she says, "is that we've had the opportunity to touch our own mortality. Once you've had that opportunity, you don't take it for granted and you have to use it as a touchstone, you have to go back to it, because it matters what you do with it."

In the first months after her accident, Alison felt as if she were looking at the world through a pane of glass. Everyone seemed so distant and disconnected. She knew she needed help, but there was no support group for people crushed by Laotian logging trucks. Slowly, with the love of friends, family, and remarkable doctors, she recovered. Today the glass pane is gone. Now she feels deeply attached to everyone around her, and she finds even the most mundane moments moving and meaningful. When she goes shopping at the vast Safeway supermarket in her neighborhood or watches a tourist bus roll by on the Golden Gate Bridge, she says, "My heart just feels so open towards these people . . . I just feel immense love for them and wish them peace, safety, and a good day."

Survivors grasp my grandmother's simple wisdom: We all hang by a thread. And like Alison Wright, they understand that we are all connected by the inevitability of adversity and universality of suffering. While our lives are as fragile as filament, the isolation and loneliness of struggle are illusions. If we band together, tethering ourselves to family, friends, and fellow survivors, we can bolster our chances of overcoming.

3. The Power of Purpose

At 8:30 AM, it's already jam-packed in the main waiting area of the renowned Dana-Farber Cancer Institute in Boston. People of all ages line up for treatments. Some cover their bald heads with scarves and hats. Others wear hospital wristbands, a giveaway that they're patients. The revolving doors seem to spin endlessly with flower deliveries. I've come here to visit a special place that focuses on both the physical and emotional needs of people with cancer.

It's called the Perini Family Survivors' Center, named after David Perini Jr., a young man who died in 1990 of Ewing's sarcoma, a rare disease of the bone and soft tissue. I knew David casually in college. He was popular, dashing, and athletic, and everyone was aware that he had cancer. When he passed away at age twenty-six after three bouts with the disease, his family wanted to honor his memory. Instead of donating to find a cure for his specific illness, they chose to fund a brand-new survivors' center. It was the first of its kind in the country. They wanted to support life and living, not death and dying.

Once a week, Eileen Perini volunteers her time at the information desk near the Dana-Farber entrance. The center is named after her family, and the memory of her son David brings her here. "I know what it's like to go through that door," she says of the constant stream of anxious, upset men and women. "I want to reach across the desk and put my arms around them." After David's death, Perini says she could have chosen to hide under her bedspread. Instead, she earned a master's degree in social work and continues to counsel students at Boston College. But her work with survivors is the most important. "It feeds me," she says. "It helps me to go on." Today the center sees 350 adult and 400 pediatric patients every year and offers a remarkable variety of medical and social services for survivors and their families. It also promotes research into new, less toxic treatments. Around the world, the Perini Center is a model for survivorship programs.

As we stroll through the bustling lobby of the Dana-Farber Institute, Perini tells me she has found her mission in life. This work is also her way of holding on to the memory of a vibrant son who lived fully and always made people feel better about themselves. "I do not want David to be gone," she says. Her blue eyes well up. "If I stopped doing this, I would let go of David," she goes on. "I cannot dishonor him by giving up." When she sees the toddlers and other pediatric patients arriving at the center for help, Perini feels the presence of her son. It's a tragedy that he never had time to marry or become a dad. But now, she believes, he is the guardian angel of so many little ones. "All those kids are his children," she says.

Generosity and selflessness are among the most surprising survival tools that I encountered in my travels. Time and again, I met men and women committed to helping others cope with misfortune. Without question, empathy, purpose, and a calling greater than oneself are among the most powerful life preservers of all.

4. Strength in Numbers

Once a month, the elderly men gather in a hotel in Ontario, California, at the base of the San Gabriel Mountains. On this crisp winter day, Cucamonga Peak is dusted with snow, and the old guys arrive bundled up in windbreakers and coats bearing the insignia of their regiments. They shed their gear and warm up in a hospitality room. They're all wearing similar uniforms: Hawaiian shirts, white pants, and caps. These are the remaining members of a proud but dwindling national organization. This is Chapter 9 of the Pearl Harbor Survivors Association.

They gather to swap stories, eat lunch, and look after one another. Survivors know there is strength in numbers, and these men hold on to each other just as they cling to their memories of December 7, 1941. On that day of infamy, the Japanese sneak attack killed 2,338 service members and civilians and sank or damaged twenty-one warships. Many of the survivors endured even more adversity. Mal Middlesworth, president of the national Pearl Harbor Survivors Association, was an eighteen-year-old on the heavy cruiser *San Francisco* when the Japanese attacked. Later as a marine, he fought in the Battle of Guadalcanal. When he returned home, he married his high school sweetheart Jojean. She often joins him at these monthly meetings.

Middlesworth is eighty-six years old now, and he marvels how he's survived all these years. He was born prematurely in the back of a railroad restaurant in Cowden, Illinois. His twin brother died at birth, and Mal weighed only two and a half pounds. They put him in a shoe box and said he'd never live to cry. "Somebody has looked out for me all these years," he says with awe. He can't even count how many times torpedoes, bullets, and bayonets narrowly missed him during the war. "I shouldn't have lived," he says. "There's always somebody helping me. I'm not that capable."

Middlesworth tells me about a recent trip to see a Pearl Harbor survivor who had suffered a stroke. It was very distressing to go to the rehab center where his buddy was in such rough shape. "They're some of the best friends I ever had," he says of his fellow survivors. But he knows the reality. There are 3.2 million World War II veterans in the United States, and around one thousand die every day. Mal estimates that only six thousand Pearl Harbor survivors are still alive. "We're losing 'em fast," he says.*

Teddy Roosevelt was a seventeen-year-old seaman aboard the battleship *Utah* when a Japanese torpedo struck at 8:01 AM. The ship sank in eight minutes. The third cousin of the twenty-sixth president, Roosevelt went on to fight in fourteen major battles in the Pacific. He's eighty-four years old and he's had a heart attack so he can't lift heavy things anymore, but he tells his cardiologist that he's determined to live to the age of ninety-two. He's very clear about his goal: at least seven more years. "I want to be the last one," he says with a twinkle. "I want to be the last survivor still breathing." I listen to his words and study Roosevelt's bright blue eyes. He's an elderly man with a heart problem but he's the personification of survival. He *wants* to live.

5. Do Not Waste a Breath

Working on this book, I was often asked if it was disturbing to spend so much time thinking about the cement truck around the corner or if it was depressing to meet so many people slammed by life. The answer to both questions is an emphatic *no*. The experience never once left me anxious or alarmed. Rather, it made me more realistic. To borrow from the military, I've got more situational awareness now. I'm more attuned to the possibility that things go wrong. I'm more alert to my blind spots and gorillas in our midst. And I'm a little more prepared. That's not to say I would fare well if you dropped me in the woods with

* "We're getting about as extinct as the dodo bird," Middlesworth told *The New York Times* at the association's final reunion in Pearl Harbor, Hawaii, in December 2006. "The way it's going, our next national convention here we could hold in a phone booth."

only one matchstick and a fishing hook. It just means that my outlook has changed. I'm more vigilant, but not to the point of distraction. To the extent that I exert any control, I know I can improve my chances.

Far from dispiriting, my encounters with survivors have been both inspiring and humbling. They may look like ordinary people, but they have accomplished extraordinary feats. When my next crisis comes, I hope to summon a fraction of their strength. Above all, I'm aware of an invisible fellowship that I never recognized before. If you go looking, you'll find members of the Survivors Club literally everywhere. You'll meet men and women of all ages struggling with their own Andes. This invisible society has chapters everywhere on earth, and its roster is always expanding. If you're open to the possibilities, you'll encounter fellow survivors who may not share your specific challenge but are ready to lend a shoulder and a hand. When you listen to them, you'll hear a common refrain. Nando Parrado captures it best with his admonition to breathe, breathe again, and cherish that you are alive. "After all these years, this is still the best advice I can give you," he writes. "Savor your existence. Live every moment. Do not waste a breath."

One breath. Then another. And another. Inhaling and exhaling. On a desolate glacier, in a hospital room, or in your own home, making the most of every breath elevates mere subsistence to real existence. Whether you reach the age of twenty, fifty, eighty, or beyond, this imperative to live fully is the most important and yet elusive lesson of the Survivors Club.

www.TheSurvivorsClub.org

As you finish this book, I invite you to visit www.TheSurvivorsClub.org, a new Web site to help people in crisis live longer and better lives. As you'll see, it's an online meeting place and information center for anyone interested in survival stories or beating the odds. In the parlance of the Internet, its mission is to connect you with tools, content, and community to cope with crisis. In other words, you'll find guides and resources to help you make the best possible decisions. You'll be able to share your own survival story and build a private support group of friends and family. And you can learn from other survivors who have already handled your precise challenge.

In one section of the site, I've included some extra information about this book. You'll discover more about the survivors in these pages and see their pictures and videos. You can print out a reading guide for book groups and find links for a number of survivorship resources and worthy causes, including those that are supported by some of the men and women in this book.

APPENDIX A

The Science of Falling Cats (and Babies)

No one knows why Leo fell out the window. He might have been chasing a bird, or he might have just fallen asleep and rolled off the sill. Whatever the reason, Leo fell forty-three floors from a New York City skyscraper. Around the thirty-seventh floor, he reached sixty miles per hour, terminal velocity for falling felines. A few seconds later, he hit the ground. Lucky for Leo, cats possess the innate ability to right themselves and align their bodies in an optimum landing position with legs stretched out, stomach flat, and head tilted back. Leo survived his forty-three-story fall with a bloody nose and one missing tooth. His lungs had collapsed—the most serious injury—but he had no broken bones. Two and half days later, he returned to his perch high above Manhattan. One life down, eight to go.

Everyone knows cats fall on their feet from any height. Perhaps less well recognized is the fact that cats often fall from tall buildings. There's even a name for the phenomenon: feline high-rise syndrome. Over the years, the science of falling cats—feline pesematology—has produced some surprising results. In general, the farther a cat falls, the fewer its injuries. In 1987, veterinarian Michael Garvey presided over a study comparing the falls of 132 cats in New York City. Nine out of ten survived. Falls below five stories weren't very damaging, mostly because velocity stayed reasonably low. In these shorter falls, cats tended to hit the ground with their legs extended like shock absorbers. They walked away bruised, disoriented, but okay. Falling between five and nine stories proved the most dangerous. Within this range, the cats reached terminal velocity but didn't have time to adjust their flight or landing positions. At these heights, they usually broke bones and sustained life-threatening injuries. Incredibly, cats that fell more than nine stories did the best of all. In the study, only one of twenty-two cats that fell from a high floor perished.*

"It's not that they can fly," says Garvey, the chief of internal medicine and critical care at Hickory Veterinary Hospital outside Philadelphia. After they reach terminal velocity, they spread their legs wide, increasing air resistance or aerodynamic drag. When they hit the ground, they flatten themselves to dissipate energy and absorb the shock. Other than flying squirrels, Garvey says, there is no other species so well equipped to fall so far and still survive.

Small children might be considered the other stars of free falling. A disturbing number of youngsters are seriously injured or killed every year in falls, but many survive. In March 2006, a two-year-old boy fell from a third-floor window in Providence, Rhode Island. He landed on a concrete sidewalk and sustained serious injuries, but he lived. "I have seen dozens of these cases over the years and I've never seen a child die," said Deputy Police Chief Paul J. Kennedy. "It's really incredible."

In fact, it's actually not so incredible when you consider a baby's body, according to Dr. Richard Snyder, the expert on impact forces you met at the end of chapter 3. Infant bones are "relatively flexible (and not rigid or brittle)," he wrote in his classic study of free falling, and their higher percentage of body fat offers "greater protection to internal injuries." Still, some falls from great heights remain a mystery. In October 1993, for instance, five-year-old Paul Rosen was reaching for a toy when he fell out the seventh-story window of his family's New York apartment onto a concrete courtyard below. In the ambulance on the way to New York Hospital, Paul asked emergency workers if he would get a Band-Aid. "Paul is a strong, brave child who thought he could fly," his mother Christine told reporters at a hospital news conference. Doctors were baffled how he survived without any injuries or broken bones. They suspected that he landed on his bottom and back, cushioning the fall. But paramedic Raymond Bonner offered a different perspective: "It's like the angels caught him."

APPENDIX B

The Arithmetic of Dying Too Soon

In eulogies and remembrances, you've heard these phrases: *He died too soon* or *Her life was cut short*. The expressions are familiar, but it turns out scientists actually quantify this concept with a statistic known as YPLL (pronounced *yipple*), a measurement of Years of Potential Life Lost. Using an average life expectancy of, say, eighty-five years, YPLL tells us how many years of future life are erased when people die too soon. For instance, when my father collapsed and died of a brain hemorrhage at age sixty-four, he was robbed of twenty-one years of potential life. It was a terrible loss for our family, but YPLL places his passing into a much larger perspective. If you add up *all* the years of potential life lost in the United States every year to cerebrovascular causes like strokes and brain bleeds, you get a YPLL of 1.14 million years. Now consider cancer. The YPLL for so-called malignant neoplasms is 8.35 million years. For heart disease, it's 6.25 million years. For "unintentional injury"—meaning accidents—the YPLL is 4.11 million years. For homicide, it's 949,607 lost years. Now add up all the causes of death in the United States—cancer, heart disease, diabetes, suicide, accidents, everything—and you get a staggering YPLL of 34.55 million years.*

* YPLL helps doctors, scientists, and policy makers understand the relative impact of disease and other causes of death and the effectiveness of programs aimed to prevent them.

NOTES

This book is based primarily on my conversations with survivors and experts around the world. Some interviews, noted below, were conducted by my research associate, Bridget Samburg. It wasn't easy selecting which survivors to include in the final draft of this book. For every person mentioned in these pages, many others shared stories and wisdom that were equally, if not more, deserving, compelling, and illuminating. In the end, I tried to choose a mixture of people who reflect the remarkable range and depth of the survivor experience.

A variety of excellent books also influenced my writing, and I'm grateful to their authors, some of whom I interviewed for additional perspective. The list includes: Laurence Gonzalez, *Deep Survival* (New York: Norton, 2003); Dr. Kenneth Kamler, *Surviving the Extremes* (New York: Penguin, 2005); John Leach, *Survival Psychology* (London: Palgrave, 1994); Joseph LeDoux, *The Emotional Brain* (New York: Touchstone, 1998); Al Siebert, *The Survivor Personality* (New York: Perigee, 1996); and Richard G. Tedeschi and Lawrence G. Calhoun, *Trauma & Transformation* (Thousand Oaks, California: Sage Publications,

1995). Unless listed below, I drew basic information from readily available news accounts, reference works, and Web sites.

PROLOGUE / *Brace for Impact*

The account of my Dunker training is based on visits to the Naval Survival Training Institute in Pensacola, Florida, and the Aviation Survival Training Center in Miramar, California. The study on the perils of drinking seawater comes from Kent B. Pandoff and Robert E. Burr, eds., *Medical Aspects of Harsh Environments,* volume 2, chapter 29, "Shipboard Medicine" (Dr. Terrence Riley), November 2002. For the Seahawk crash of January 26, 2007, see Steve Liewer, "Helicopter Crash That Killed Navy Crew Still a Mystery," *San Diego Union-Tribune,* December 23, 2007.

INTRODUCTION / *The Survivors Club*

For more on the concept of active passiveness, see John Leach, *Survival Psychology,* pages 166–169.

1 / *A Knitting Needle Through the Heart*

The story of the knitting needle through Ellin Klor's heart comes from several interviews with her and her trauma doctors, including David Spain and Susan Brundage. For more on crocodile hunter Steve Irwin's death, see Nadia Salemme, "Irwin's Fatal Instinct," *MX* (Australia), September 22, 2006. The quotation in the footnote about Americans at risk of natural disasters is drawn from Amanda Ripley, "Why We Don't Prepare," *Time,*

August 28, 2006. The etymology of *survivor* comes from www.etymonline.com. The idea of survival as "living fully" is based on my interview with Amy Grose, a licensed social worker and psychosocial program leader with the Perini Family Survivors' Center at the Dana-Farber Cancer Institute in Boston. For an excellent overview of co-survivorship, see Jennifer Haupt, "A New Survivors' Club," *Cure,* spring 2006. The statistics on risks to co-survivors come from the National Family Caregivers Association Web site (www.nfcacares.org), citing Elissa D. Epel, University of California–San Francisco, in *The Proceedings of the National Academy of Sciences* 101:49, December 7, 2004. For more on the danger of losing a spouse, see Dr. Nicholas A. Christakis et al., "Mortality After the Hospitalization of a Spouse," *New England Journal of Medicine* 354:7, February 16, 2006.

I drew the concept of the "new normal" from an essay by Nate Berkus, "Surviving the Tsunami," *People,* December 26, 2005. For controversy over the word *survivor,* see Gina Kolata, "In One Word, an Entire Debate on Cancer," *New York Times,* June 1, 2004; and Barbara Sills, "That's Cancer Veteran, Conqueror, or Activist to You," *Los Angeles Times,* November 12, 2007. My thinking about survivorship was greatly influenced by my interview with Dr. Fitzhugh Mullan of George Washington University. Mullan launched a survivorship revolution with his pioneering essay "Seasons of Survival: Reflections of a Physician with Cancer," *New England Journal of Medicine,* July 25, 1985, pages 270–273. I also recommend his powerful memoir, *Vital Signs: A Young Doctor's Struggle with Cancer* (New York: Farrar Straus Giroux, 1983). My writing was also influenced by a conversation with Ellen Stovall, the dynamic and inspiring former president and CEO of the National Coalition for Cancer Survivorship. Stovall was diagnosed with Hodgkin's disease in 1971 and experienced a recurrence twelve years after her initial treatment.

The account of Nando Parrado's ordeal in the Andes was based on several interviews with him. I also relied extensively on his memoir written with Vince Rause, *Miracle in the Mountains* (New York: Crown, 2006). Various quotations including his interior monologue—*Do not cry. Tears waste salt . . .* — and his father's question—"How did you survive?"—come from the book. For Parrado's encounter in Utah with the distraught woman who accidentally killed her child, I relied on his book and our conversations. For additional details about the ordeal in the Andes, I drew on the classic by Piers Paul Read, *Alive* (New York: Avon, 1974). Terry Anderson's story is based on my interview. Some details and quotations come from his book *Den of Lions* (New York: Ballantine, 1993) and articles including Bill Sloat, "The Candidate," *Plain Dealer (Cleveland),* July 11, 2004.

For the grandmother who lifted the Chevy, I relied on an interview with Angela Cavallo conducted by my research associate. See also the Associated Press, "Huge Car Lifted by Grandmother," April 13, 1982. For more on the strength limits of the human body, see Michio Ikai and Arthur H. Steinhaus, "Some Factors Modifying the Expression of Human Strength," *Journal of Applied Physiology* 16, 1961, pages 157–163.

2 / *The Statues in the Storm*

For the stages of hypothermia, I relied on handouts from the Naval Survival Training Institute in Pensacola, Florida. The account of Paul Barney's ordeal is based on my interview with him. Additional details were drawn from Eleanor Mills, "The Drowned and the Saved," *The Observer,* January 28, 1996; Esther Oxford, "I Survived . . . ," *The Independent,* September 28, 1995; James Meek, "Under Uncaring Sky . . . ," *Vancouver*

Sun, October 1, 1994; and Paul Harris, "How I Clung on to Life," *Daily Mail*, September 30, 1994.

The section on the Professor of Survival comes from my interviews with Dr. John Leach and his book *Survival Psychology.* The section on no-pull parachute fatalities comes from my interviews with James D. Griffith and Christian L. Hart and their articles, including "An Analysis of US Parachuting Fatalities: 2000–2004," *Perceptual and Motor Skills* 103:3, December 2006; and "Rise in Landing-Related Skydiving Fatalities," *Perceptual and Motor Skills* 97:2, October 2003. For John Leach's work on brainlock, I relied on our conversations and his article with Rebecca Griffith, "Restrictions in Working Memory Capacity During Parachuting: A Possible Cause of 'No Pull' Fatalities," *Applied Cognitive Psychology* 22:2, 2008, pages 147–157. The risk of dying from falling down stairs comes from Larry Laudan, *The Book of Risks* (New York: John Wiley, 1994). The section on the Stockdale paradox comes from Jim Collins, *Good to Great: Why Some Companies Make the Leap . . . and Others Don't* (New York: Collins, 2001), chapter 4. The phrase *cosmic coin toss* comes from my interview with Al Siebert and his book *The Survivor Personality.*

For the Theory of 10–80–10, I relied on my interviews with John Leach and chapter 2 of *Survival Psychology,* "Psychological Responses to a Disaster." I also recommend his articles: "Why People 'Freeze' in an Emergency: Temporal and Cognitive Constraints on Survival Responses," *Aviation, Space and Environmental Medicine* 75:6, June 2004; and "Cognitive Paralysis in an Emergency: The Role of the Supervisory Attentional System," *Aviation, Space and Environmental Medicine* 76:2, February 2005. See also his articles with Louise Ansell, "Impairment in Attentional Processing in a Field Survival Environment," *Applied Cognitive Psychology* 22:5, 2008, pages 643–652; and with S. J. Robinson et al., "The Effects of Exposure to an Acute Naturalistic Stressor on Working Memory,

State Anxiety and Salivary Cortisol Concentrations," *Stress* 11, 2008, pages 115–125.

For the section on Professor Popsicle, I relied on an interview with Gordon Giesbrecht conducted by my research associate and a variety of excellent profiles including Alisa Smith, "Meet Prof. Popsicle," *Outside,* January 2003; and Joanne Laucius, "King of Cold Put Himself on Ice to Test Hypothermia Theories," *Ottawa Citizen* (Canada), February 16, 2005. For the 1–10–1 System, see Kevin Rollason, "Professor Offers Cold-Weather Facts, Survival Tips," *Winnipeg Free Press* (Canada), October 18, 2006. For myths of hypothermia, see Jennifer Nichols, "Survival Seminar Dispels Myths," *Meadow Lake Progress* (Saskatchewan, Canada), February 18, 2007; and Lara Bradley, "The Iceman Cometh," *Sudbury Star* (Canada), November 22, 2004. For more on hypothermia and survival, I recommend Giesbrecht's book with James A. Wilderson: *Hypothermia, Frostbite and Other Cold Injuries* (Seattle: Mountaineers, 2006).

3 / *Ninety Seconds to Save Your Life*

The account of Jerry Schemmel's survival on United 232 is based on his interview with my research associate, our e-mail exchanges, and his memoir with Kevin Simpson, *Chosen to Live* (Littleton, Colorado: Victory, 1996). The paragraphs on Captain Al Haynes are drawn from my interview with him and his writing, including "Eyewitness Report: United Flight 232" from www.airdisaster.com. The Arnold Barnett section is based on my interview with him and his articles, including "Air Safety: End of the Golden Age?" *Chance* 3, February 1990; and "Measure for Measure," *Aerosafety World,* November 2007.

I owe many of the facts and ideas in this chapter to my visit to the FAA's Cabin Safety Workshop in Oklahoma City. For an excellent analysis of airplane safety and evacuations, see Barbara S. Peterson, "The Great Escape," *Condé Nast Traveler*, November 2005. The etymology of *panic* comes from www.etymonline.com The quotations from Lee Clarke come from his interview with my research associate and his essay "Panic: Myth or Reality," *Contexts* 1:3, fall 2002. The quotations from Daniel A. Johnson are drawn from his book *Just in Case: A Passenger's Guide to Airplane Safety and Survival* (New York: Plenum Press, 1984). The study on the incapacitation of flight attendants is described in Johnson's book. My thinking on evacuations and competitive behavior was shaped in many ways by my interview with Professor Helen Muir, head of human factors and director of the Institute for Safety, Risk and Reliability at Cranfield University in England.

The section on David Koch and USAir 1493 comes from my interview with him; his personal essay, "Recollections of My Survival of an Airplane Crash," February 13, 1991; and his article "Passenger's Account of Escape from Burning Boeing 737 . . . ," *Cabin Crew Safety*, Flight Safety Foundation, 28:3, May/June 1993. See also National Transportation Safety Board, "Aircraft Accident Report: Runway Collision of USAir Flight 1493 . . . ," October 22, 1991. The paragraph on Nora Marshall comes from my interview with her. The material on the best seat on the plane comes from David Noland, "Safest Seat on a Plane," *Popular Mechanics*, July 18, 2007; my conversations with aviation safety officials at the FAA and NTSB; and my interview and e-mail exchanges with Ed Galea. For the Five Row Rule and Greenwich study, see Galea et al., "A Database to Record Human Experience of Evacuation in Aviation Accidents," Civil Aviation Authority (UK), Paper 2006/01, June 2008, pages 36–41.

The footnote on terrorism, stress, and heart risk comes from E. Alison Holman et al., "Terrorism, Acute Stress, and Cardiovascular Health," *Archives of General Psychiatry* 65:1, January

2008. The mortality estimate from *The New York Times* appeared in John Tierney's column, "Living in Fear and Paying a High Cost in Heart Risk," January 15, 2008.

For the section on the woman who fell from the sky, I relied on Kate Swoger, "Sole Survivor Looks Back," *Prague Post,* February 6, 2002; "Survivors: Five Women Who Faced Certain Death . . . ," *Life,* December 1999; Eve-Ann Prentice, "Suddenly a Bomb Blew the Jet Apart," *The Times* (London), June 26, 2001; "Woman Who Survived Fall from Exploding Plane Back at Scene," Agence France-Presse, January 26, 2002; and Philip Baum, "How to Survive a Bombing at 33,000 Feet," *Aviation Security International,* November 1, 2002. For Vulovic's life today—and her quotation "If you can survive what I survived . . ."—I relied on Dan Bilefsky, "Serbia's Most Famous Survivor Fears That Recent History Will Repeat Itself," *New York Times*, April 26, 2008.

For the section on Richard Snyder, I relied on my interview with him and his article, "Human Survivability of Extreme Impacts in Free-Fall," Civil Aeromedical Research Institute, Federal Aviation Agency, CARI Report 63-15, August 1963. For more on free falls, see the Free Fall Research Page: www.greenharbor.com/fffolder/ffresearch.html. See also Jim Hamilton, *Long-Fall Survival: Analysis of Collected Accounts* (Marshfield, Massachusetts: Green Harbor Publications, 2006).

4 / *The Organ Recital*

This chapter is based largely on my visit to the trauma department at Stanford University Medical Center, my attendance at the Trauma Survivor Reunion, and interviews with Dr. David Spain and Dr. Susan Brundage. The survival stories come from conversations with Katharine Decker Johnson, Ricky Bunch,

Steve Herrera, and Gary McCane Jr. For more on the Survivor Gene, see Simon V. Baudouin, David Saunders, et al., "Mitochondrial DNA and Survival After Sepsis: A Prospective Study," *The Lancet* 366:9503, December 17, 2005. Information on salvage rates at accident scenes comes from my interview with Dr. Bryan Bledsoe.

The section on the Golden Gate Bridge comes from my interviews with Ken Holmes and Kevin Hines, and from a number of excellent articles, including Tad Friend, "Letter from California: Jumpers," *The New Yorker,* October 13, 2003; and a seven-part series in the *San Francisco Chronicle* titled *Lethal Beauty.* The *Chronicle* series included an insightful profile of Kevin Hines written by Mike Weiss, "A Survivor's Story," November 1, 2005. The "loaded gun" quotation comes from Heidi Benson's article in the *Chronicle* series, "Saving a Life," November 5, 2005. For the success rate of suicide techniques, see Geo Stone, *Suicide and Attempted Suicide: Methods and Consequences* in Edward Guthmann's article in the *Chronicle* series, "The Allure," October 30, 2005.

For the section on the best place to have a heart attack, the quotation from Thurman Austin comes from his interview with my research associate. For the history of casinos and defibrillators, I drew on Kevin Helliker's excellent article, "Beating the Odds," *Wall Street Journal,* January 28, 2006; and Pat King, "Local Casinos Using Heart Devices to Help Save the Lives of Their Customers," *Las Vegas Business Press,* June 12, 2000. For background information on cardiac arrest salvage rates, I relied on my interview with Dr. Bryan Bledsoe. For the definitive study on cardiac arrest and casinos, see Terence D. Valenzuela et al., "Outcomes of Rapid Defibrillation by Security Officers After Cardiac Arrest in Casinos," *New England Journal of Medicine* 343:17, October 26, 2000. For more on deadly defibrillation delays, see Paul S. Chan et al., "Delayed Time to Defibrillation After In-Hospital Cardiac Arrest," *New England Journal of Medicine* 358:1, January 3, 2008; and Mary Ann Peberdy et al., "Survival from In-Hospital Cardiac Arrest During Nights and Weekends," *Journal of the American Medical Association* 299:7, February 20, 2008.

5 / *The Supersonic Man*

The description of g-forces and everyday activities comes from a famous study by Murray E. Allen et al., "Acceleration Perturbations of Daily Living," *Spine* 19:11, 1994. The account of Brian Udell's ejection is based on my interviews with him. See also Technical Sergeant Timothy P. Barela, "Back in the Saddle," *Airman,* April 1997; and "Capt. Survived Horrific Ejection," *Morning Star (Wilmington),* November 10, 1997. For SR-71 ejections above sixty thousand feet, see Richard H. Graham, *SR-71 Revealed: The Inside Story* (Osceola, Wisconsin: Zenith Press, 1996). Details and quotations from Udell's visit to the ejection seat factory in Phoenix are drawn from Brahm Resnik, "Plant Perfects Ejection Seats," *Arizona Republic,* December 9, 2001.

The sections on Ken Kamler, the doctor of extremes, are based on my interviews with him and his book *Surviving the Extremes*. For more on La Rinconada in Peru, see *High Altitude Medicine & Biology,* November 2002. I first heard the phrase *let go and let God* from Tom Lutyens, the trainer and contractor with the Air Force Survival School in Spokane, Washington. The description of the magic numbers of staying alive is based on interviews with Lutyens and Technical Sergeant Joshua Anderson, an air force survival specialist.

6 / *Rescued from the Lion's Jaws*

The account of Anne Hjelle's mountain lion attack is drawn from my interviews with her. For the masticatory power of Hercules the alligator, see G. M. Erickson et al., "The Ontogeny of Bite-Force Performance in the American Alligator," *Journal of Zoology* 260, pages 317–327. For *T. rex* and other creatures, see G. M. Erickson et al., "Bite-Force Estimation for *Tyrannosaurus Rex* from Tooth-Marked Bones," *Nature* 382, August 22, 1996, pages 706–708. Debi Nicholls's quotation—"I'm never letting go . . . "—comes from the Associated Press, January 9, 2004.

The danger of beach apple trees comes from Ray E. Smith and D. Shiras Jarvis, eds., *How to Survive on Land and Sea,* fourth edition (Annapolis: Naval Institute Press, 1984). The story of Stanley Praimnath's deliverance from the World Trade Center is based on my interview with him and his book with William Hennessy, *Plucked from the Fire* (Pittsburgh: RoseDog Books, 2004).

The section on the healing power of faith is drawn from my interview with Dr. Harold G. Koenig, the leading thinker and researcher in this field, and from his extensive writing, including *The Healing Power of Faith: How Belief and Prayer Can Help You Triumph Over Disease* (New York: Touchstone, 1999). For Dr. Koenig's remarkable personal story, see his book with Greg Lewis, *The Healing Connection: The Story of a Physician's Search for the Link Between Faith and Health* (Philadelphia: Templeton, 2000). For the seven-year study, see R. A. Hummer et al., "Religious Involvement and US Adult Mortality, *Demography* 36, 1999, pages 273–285. The section on religious struggle comes from my interview with Kenneth I. Pargament and his article "Religious Struggle as a Predictor of Mortality Among Medically Ill Elderly Patients," *Archives of Internal Medicine* 161, August 13/27, 2001.

The Reverend Lin Barnett's story comes from my interview with him and from Brian Dugger, "House of Healing, House of Prayer," Scripps Howard News Service, August 23, 2000. For the study that found *no* benefit to intercessory prayer, see H. Benson et al., "Study of the Therapeutic Benefits of Intercessory Prayer in Cardiac Bypass Patients," *American Heart Journal,* April 2006. For the meta-study on intercessory prayer, I relied on my interview with David R. Hodge and his article "A Systematic Review of the Empirical Literature on Intercessory Prayer," *Research on Social Work Practice,* March 2007.

The section on the sanctuary in Chimayo is based on my interview with Father Casmiro Roca and Kate McGraw's article, "The Father of Chimayo," *Albuquerque Journal,* December 21, 2003. The quotations from Carl Sagan come from *The Demon-Haunted World: Science as a Candle in the Dark* (New York: Ballantine, 1996). The section on spontaneous remission is based on my research associate's interview with Caryle Hirshberg and her book with Ian Marc Barasch, *Remarkable Recovery: What Extraordinary Healings Can Teach Us About Getting Well and Staying Well* (New York: Riverhead, 1995).

The section on the God Helmet is based on my interview with Dr. Michael Persinger and various articles including Jack Hitt's excellent "This Is Your Brain on God," *Wired,* November 1999; and Ian Cotton, "Dr. Persinger's God Machine," *The Independent,* July 2, 1995. For the Swedish rebuttal, see Pehr Granqvist et al., "Sensed Presence and Mystical Experiences Are Predicted by Suggestibility . . . ," *Neuroscience Letters* 379:1, April 29, 2005.

7 / *The Dancer and the Angel of Death*

Edie Eger's story of survival in Auschwitz is based on my interviews with her and John M. Glionna, "Taming the Demons,"

Los Angeles Times, January 12, 1992; and Maria Hagedorn, "Hope Amidst the Ashes of the Holocaust," *Faith and Friends*, January 2002. The quotations from Charles Krauthammer come from his article "Holocaust," *Time*, May 3, 1993. Quotations from Bruno Bettelheim come from *Surviving and Other Essays* (New York: Vintage, 1979). The first quotation from Elie Wiesel—"Those who have not lived . . ."—appears in Bettelheim's *Surviving* on page 96. The quotations from Terrence Des Pres come from *The Survivor: An Anatomy of Life in the Death Camps* (Oxford: Oxford University Press, 1976). For the portion on lighter physical features, I relied on my research associate's interview with Peter Suedfeld and his article, "Lethal Stereotypes: Hair and Eye Color as Survival Characteristics During the Holocaust," *Journal of Applied Social Psychology*, 2002, pages 2368–2376. The words of Primo Levi come from *The Drowned and the Saved* (New York: Simon & Schuster, 1988). See also Levi's classic: *Survival in Auschwitz* (New York: Touchstone, 1996). The material from Viktor E. Frankl comes from his classic work *Man's Search for Meaning* (Boston: Beacon, 2006).

The section on Elie Wiesel is based on my interview with him and his devastating memoir *Night* (New York: Hill and Wang, 2006). The portion on William Helmreich's work is based on my interviews with him and his book *Against All Odds* (New Brunswick, New Jersey: Transaction, 1996). The interpretation of the Hebrew word *mazal* comes from my conversation with Helmreich. On the role of luck in Holocaust survival, I relied extensively on Henry Friedlander and Sybil Milton's essay "Surviving" in Alex Grobman and Daniel Landes, eds., *Genocide: Critical Issues of the Holocaust* (Los Angeles: Simon Wiesenthal Center; Dallas: Rossell Books, 1983).

The account of Rachel Yehuda's work is based on my interviews with her and e-mail exchanges. See also Rachel Yehuda et al., "Low Urinary Cortisol Excretion in Holocaust Survivors with Posttraumatic Stress Disorder," *American Journal of Psychiatry* 152:7, July 1995; Yehuda et al., "Parental Posttraumatic Stress Disorder as a Vulnerability Factor for Low Cortisol Trait in Offspring of Holocaust Survivors," *Archives of General Psychiatry* 64:9, September 2007; Yehuda et al., "Ten-Year Follow-Up Study of Cortisol Levels in Aging Holocaust Survivors With and Without PTSD," *Journal of Traumatic Studies* 20:4, August 2007. For the 9/11 study, see Rachel Yehuda et al., "Transgenerational Effects of Posttraumatic Stress Disorder in Babies of Mothers Exposed to the World Trade Center Attacks During Pregnancy," *Journal of Clinical Endocrinology and Metabolism,* May 3, 2005.

8 / *The Science of Luck*

I am greatly indebted to Professor Richard Wiseman for this chapter, especially his book *The Luck Factor,* the main source for many quotations and descriptions of his fascinating and unusual research. I was unable to interview Professor Wiseman, but he answered questions via e-mail, and his responses are used throughout. For background and quotations, I also drew on articles about Dr. Wiseman's work including Daniel H. Pink, "How to Make Your Own Luck," *Fast Company*, June 2003.

The account of the gorilla study comes from my interview with Daniel J. Simons. See also his article with Christopher F. Chabris, "Gorillas in Our Midst: Sustained Inattentional Blindness for Dynamic Events," *Perception* 28, 1999, pages 1059–1074. The *opportunityisnowhere* test comes from Dr. W. Gifford-Jones, "WWI Pilots Beat the Odds: What's Their Secret to Survival During and After War?" *Toronto Sun* (Canada), October 27, 1994. The section on the Luck Formula comes from my e-mail exchanges with Nicholas Rescher and his book *Luck: The Brilliant Randomness of Everyday Life* (Pittsburgh: University of Pittsburgh, 1995).

The serendipity story and quotations come from Al Siebert's *The Survivor Personality*. The section on Cindy Roper comes from my interview with her and an article by Ben Swan, "Living with Plague," *New Mexican (Santa Fe),* August 18, 2007. For more on bronze medals versus silver medals, see V. H. Medvec, S. F. Madey, and T. Gilovich, "When Less Is More: Counterfactual Thinking and Satisfaction Among Olympic Medalists," *Journal of Personality and Social Psychology,* 1995.

The section on Samantha Dunn is based on my interviews with her and her excellent memoir *Not By Accident* (New York: Henry Holt, 2002). I also drew on her article "A Life of Accidents, A Quest for Answers," *Los Angeles Times,* May 6, 2002. Sigmund Freud's quotation appears in Dunn's book on page 139. For accident proneness, I relied on my interview with Ellen Visser and her article, "Accident Proneness, Does It Exist? A Review and Meta-Analysis," *Accident Analysis and Prevention* 39, 2007, pages 556–564. For accidents versus injidents, see *British Medical Journal* 322, June 2, 2001, pages 1320–1321.

The section on Luck School is drawn from Wiseman's book. For more on Technical Asset Management, the company that increased its luck, see Martin Plimmer and Brian King, *Beyond Coincidence: Amazing Stories of Coincidence and the Mystery and Mathematics Behind Them* (New York: Thomas Dunne Books, 2006).

The section on lefties versus righties is based on my interview with Stanley Coren and his book *The Left-Hander Syndrome* (New York: Vintage, 1993). The etymology of *sinister* comes from www.etymonline.com.

9 / *Hug the Monster*

I'm very grateful to Dr. Joseph LeDoux for his significant contributions to this chapter. The accounts of his life and work are based on my interviews with him and his book *The Emotional Brain*. For more on the Amygdaloids, see Roja Heydarpour, "A Band of Scientists Who Really Are a Band," *New York Times*, March 6, 2007; and Jonathan Cott and Karen Rester, "Joseph LeDoux's Heavy Mental," Salon.com, July 25, 2007. For more on erasing traumatic memories with drugs, see Michael Behar and Saba Berhie, "Paging Dr. Fear," *Popular Science* 272:1, January 2008; Valerie Doyere et al., "Synapse-Specific Reconsolidation of Distinct Fear Memories in Lateral Amygdala," *Nature Neuroscience* 10, March 11, 2007, pages 414–416: and R. K. Pitman et al., "Pilot Study of Secondary Prevention of Posttraumatic Stress Disorder with Propranolol," *Biological Psychiatry* 51:2, January 15, 2002. The section on 9/11 and active coping comes from LeDoux and Jack M. Gorman, "A Call to Action: Overcoming Anxiety Through Active Coping," *American Journal of Psychiatry*, December 2001.

Tim Sears's survival story is based on his interview with my research associate and various newspaper accounts, including Kevin Moran, "Cruise Passenger Saved After 17 Hours in the Gulf," *Houston Chronicle*, April 18, 2003, and "Survivor Says 'Will to Live' Saved Him," April 19, 2003; James McCurtis, "Lansing Man Survives Fall from Cruise Ship," *Lansing State Journal*, April 19, 2003; and Marney Rich Keenan, "Fall from Cruise Ship Makes Tim Sears a Media Darling," *Detroit News*, April 26, 2003.

The section on the smell of fear is based on my interview with Denise Chen. See also Denise Chen and Jeannette Haviland-Jones, "Human Olfactory Communication of Emotion," *Perceptual and Motor Skills* 91, 2000, pages 771–781; and Denise Chen et al., "Chemosignals of Fear Enhance Cognitive Performance in Humans," *Chemical Senses* 31:5, March 9, 2006. Adair Rowland's story is based on her interview with my research associate and our extensive e-mail exchanges.

Antonio Hansell's encounter with a brown bear is based on

news accounts in Boston, including Anne Saunders, "Mass. Camper Dies in NH After Bear Scare," Associated Press, August 5, 2004. The section on the Baskerville effect is based on my interview with Dr. David Phillips and his article with others, "The Hound of the Baskervilles Effect: Natural Experiment on the Influence of Psychological Stress on Timing of Death," *British Medical Journal* 323:7327, December 22, 2001. The passages on Dr. Martin Samuels are based on his interview with my research associate and Anupretta Das, "Scared to Death," *Boston Globe,* August 6, 2006. Dr. Samuels's time bomb quotation comes from ABCNews.com, "Being 'Scared to Death' Can Kill," October 30, 2006. For Dr. Walter Cannon's classic study see "'Voodoo' Death," *American Anthropologist* 44, 1942, pages 169–181. Laurence Gonzalez's quotation on fear comes from *Deep Survival.*

The section on who gets lost is based on my interview with Ken Hill and the book that he edited: *Lost Person Behaviour* (Ottawa: National Search and Rescue Secretariat, December 1999). For details of the Andrew Warburton story, I drew on Dean Beeby, *Deadly Frontiers: Disaster and Rescue on Canada's Atlantic Seaboard* (Fredericton: Goose Lane, 2001).

10 / *Too Mean to Die*

Trisha Meili's story is based on my interviews with her and her compelling memoir, *I Am the Central Park Jogger* (New York: Scribner, 2003). Quotations about the will to live come from Department of the Army Field Manual FM 21-76, *Survival* (Washington: Department of the Army, 1970). The account of Dr. Steven Greer's work is based on my interview with him and his articles, including "Psychological Response to Breast Cancer: Effect on Outcome," *The Lancet*, October 13, 1979; "Influence of Psychological Response on Survival in Breast

Cancer: A Population-Based Cohort Study," *The Lancet*, October 16, 1999; and "Fighting Spirit in Patients with Cancer," *The Lancet*, March 4, 2000. The italics are mine in the quotation: "It is not what may be *added in* by fighting but what is *taken away* . . ." The quantity versus quality of life distinction comes from my interview with Dr. Kevin Stein, director of quality of life research with the American Cancer Society.

The section on Dr. Samuel Gruber is based on my interviews with him. The account of Mary Ward's attack comes from my interview with Dave LaBahn and various news accounts, including Rene Lynch, "Rape Suspect's Murder Trial Tests Legal Limits," *Los Angeles Times,* January 24, 1994; Timothy Appleby, "Will to Live on Trial in LA Crime," *Globe and Mail* (Toronto), January 31, 1994; Rene Lynch, "Witness Says Rape Changed Victim," *Los Angeles Times,* February 3, 1994; and Byron MacWilliams, "Did Rapist Kill Woman's Spirit?" *Orange County Register,* March 11, 1994.

The portions on the interior cingulate gyrus come from Dr. Kenneth Kamler, *Surviving the Extremes*. For Alice Trillin's article, see "Of Dragons and Garden Peas: A Cancer Patient Talks to Doctors," *New England Journal of Medicine* 304:12, March 19, 1981. The section on postponing death comes from my interview with David Phillips and his article with Kenneth A. Feldman, "A Dip in Deaths Before Ceremonial Occasions: Some New Relationships Between Social Integration and Mortality," *American Sociological Review,* December 1973, page 678. See also Phillips et al., "The Birthday: Lifeline or Deadline?" *Psychosomatic Medicine,* 1992. For the rebuttal, see Dr. Donn C. Young, "Does Death Take a Holiday?" *Journal of the American Medical Association,* December 2004.

The section on Jeanne Louise Calment, the woman who lived 122 years, is based on news accounts and obituaries. Her quotation—"I dream, I think . . ."—comes from Amy Barrett, "World's Oldest Rapper Celebrates 121st Birthday," Associated

Press, February 21, 1996. The quotation—"Always keep your smile . . . I think I will die laughing"—comes from Reuters, "At 120 Years Plus 238 Days, It's a Record," *Chicago Tribune,* October 16, 1995. The section on predicting longevity is based on my research associate's interview with Dr. James Vaupel. See also, Gina Kolata, "Live Long? Die Young? Answer Isn't Just in Genes," *New York Times*, August 31, 2006.

11 / *The Resilience Gene*

Cindi Broaddus's story is based on my interview with her and her poignant memoir with Kimberly Lohman Suiters, *A Random Act* (New York: Morrow, 2006). For the groundbreaking study on the Resilience Gene, see Avshalom Caspi et al., "Influence of Life Stress on Depression: Moderation by a Polymorphism in the 5-HTT Gene," *Science* 301:5631, July 18, 2003. For the Dunedin study, I relied on my interview with Richie Poulton. The genetics of resilience are drawn from my interviews with Ian Craig. For an excellent examination of resilience and G x E, see Emily Bazelon, "A Question of Resilience," *New York Times Magazine,* April 30, 2006.

The section on Jeff Zucker is based on my interview with him and Patricia Sellers, "Life Imitates TV," *Fortune,* April 30, 2007. The Resilience Prescription comes from my interviews with Dr. Andy Morgan of the Yale School of Medicine and Dr. Dennis Charney of Mount Sinai School of Medicine.

The section on killer initials is based on my interview with Dr. Nicholas Christenfeld and his article with David P. Phillips and Laura M. Glynn, "What's in a Name: Mortality and the Power of Symbols," *Journal of Psychosomatic Research* 47:3, 1999. The rebuttal can be found in Stilian Morrison and Gary Smith, "Monogrammic Determinism?" *Psychosomatic Medicine* 67, 2005, pages 820–824.

12 / *What Does Not Kill Me*

The section on posttraumatic growth stems from my interviews with Lawrence Calhoun and his books with Richard G. Tedeschi, *Trauma and Transformation: Growing in the Aftermath of Suffering* and *Facilitating Posttraumatic Growth: A Clinician's Guide* (Mahwah, New Jersey: Lawrence Erlbaum, 1999). The section on Cassi Moore is based on my interview with her.

The section on POWs is based on my visit to the Robert E. Mitchell Center for Prisoner of War Studies in Pensacola, Florida, and interviews with Dr. Bob Hain, Dr. Jeff Moore, Dr. Robert Mitchell, and Dr. Fred Wells. The study of American prisoners in the Korean War is cited in Adriana Feder et al., "Posttraumatic Growth in Former Vietnam Prisoners of War," paper submitted for publication to *American Journal of Psychiatry,* page 4. The section on self-healing is based on my interview with Richard Mollica and his book *Healing Invisible Wounds* (Orlando: Harcourt, 2006).

The survival secrets of the Methuselah Tree are drawn from my research associate's interview with Thomas Swetnam, our e-mail exchanges, and my conversations with Tom Harlan of the University of Arizona and LeRoy Johnson. I also drew from Carl T. Hall's superb profile of the ancient tree, "Staying Alive," *San Francisco Chronicle,* August 23, 1998, and *Nova*'s outstanding documentary and Web site, "Methuselah Tree," December 11, 2001.

13 / *The Survivor Profiler*

The descriptions of the Air Force Survival School are based on my visit to the 336th Training Group at Fairchild Air Force Base in Spokane, Washington, and an excursion into a training area in the Kaniksu National Forest. The section on the development of the Survivor Profiler is based on my conversations and work with Dr. Courtney McCashland and her team at TalentMine in Lincoln, Nebraska. Dr. Al Siebert's quotations come from my interview with him and his book *The Survivor Personality.* For a glimpse of Siebert's early work, see his article "The Survivor Type Person," *Mensa Journal,* March 1969.

15 / *Your Survivor Tool Kit*

I found the Daruma doll reference in Richard Wiseman's *The Luck Factor.* Harold Koenig's quotations on faith come from *The Healing Power of Faith* and *The Healing Connection.* Dr. Jerome Groopman's quotations on hope come from *The Anatomy of Hope* (New York: Random House, 2004). For a complete list of MacGyverisms, see www.macgyveronline.org. Laurence Gonzalez's quotations come from *Deep Survival.* The quotations from Gavin de Becker come from his book *The Gift of Fear* (New York: Dell, 1997).

AFTERWORD / *How to Eat an Elephant*

I learned the phrase *how to eat an elephant* from Tom Lutyens, the air force survival expert. The section on Alison Wright is based on my interview with her and her award-winning *Outside* magazine articles "If I Can Only Breathe," May 2001, and "The Life That Almost Wasn't," February 2005. I enthusiastically recommend her memoir *Learning to Breathe: One Woman's Journey of Spirit and Survival* (New York: Viking Press, 2008). The section on Eileen Perini is based on my interview with her and visit to the Perini Family Survivors' Center at the Dana-Farber Cancer Institute in Boston. The section on the Pearl Harbor Survivors Association is based on interviews with Mal Middlesworth and Teddy Roosevelt and my visit to a monthly meeting of Chapter 9 in Ontario, California. The dodo bird quotation comes from Jesse McKinley, "Pearl Harbor Veterans Rally for a Final Hawaii Reunion," *New York Times*, December 8, 2006. The final quotation from Nando Parrado is drawn from the last page of *Miracle in the Andes*.

APPENDIX A / *The Science of Falling Cats (and Babies)*

The science of falling cats comes from Dr. Michael Garvey's interview with my research associate and Wayne O. Whitney and Cheryl J. Mehlhaff, "High-Rise Syndrome in Cats," *Journal of the American Veterinary Medical Association* 191:11, December 1, 1987. For falling cats in Croatia, see D. Vnuk et al., "Feline High-Rise Syndrome: 119 Cases (1998–2001)," *Journal of Feline Medicine & Surgery* 6:5, December 31, 2003. The quotation from Deputy Police Chief Paul Kennedy comes from Steve Peoples, "Baby Falls 30 Feet from Third-Floor Window," *Providence Journal*, March 22, 2006. The "angels" quotation from paramedic Raymond Bonner comes from Rick Hampson, "Paul's Fall: Incredible!" Associated Press, October 25, 1993. See also Charisse Jones, "How 5-Year-Old Boy Survived 7-Story Fall Baffles Medical Experts," *New York Times*, October 26, 1993.

APPENDIX B / *The Arithmetic of Dying to Soon*

This section is based on information from the National Center for Injury Prevention and Control and its 2005 report on Years of Potential Life Lost in the United States. For more detailed information on YPLL reports, go to the Centers for Disease Control Web site: www.cdc.org.

ACKNOWLEDGMENTS

I offer my deepest gratitude and respect to the survivors whose words and wisdom make up the heart and soul of this book. I interviewed many other men and women whose survival experiences and perspectives enriched my writing, and you will find more of their absorbing stories on the Survivors Club Web site. This book also could not exist without the tutelage of the experts who generously shared their time and research and who reviewed many of the chapters. I am indebted to all of them.

Captain Donnie "Spike" Plombon arranged my visit to the Aviation Survival Training Center in Miramar, California. For sharing their expertise in dunking and drown-proofing, I salute him and Lieutenant Commander Rebecca "Sparky" Bates, Lieutenant Commander Russell "Crazy Juice" Linderman, and Hospital Corpsman First Class Ralph D. Nay.

Dr. Bob Hain welcomed me to the Robert E. Mitchell Center for Prisoner of War Studies in Pensacola, Florida. I thank him and Dr. Jeff Moore, Dr. Robert Mitchell, and Dr. Fred Wells.

Technical Sergeant Joshua Anderson and Tom Lutyens escorted me on an excellent adventure in the woods of the Air Force Survival School near the Canadian border with Washington State. Gratitude also to Colonel Jeffrey White, Chief Master Sergeant Bill Welch, Major Glen Fisher, and the rest of the 336th Training Group at Fairchild Air Force Base.

Dr. Mac McLean orchestrated my visit to the FAA's Cabin Safety Workshop in Oklahoma City. I'm thankful to him, Rick DeWeese, Cynthia Corbett, and David Palmerton for their hospitality and expertise.

Dr. David Spain invited me to the Trauma Survivor Reunion at Stanford. I'm indebted to him and Dr. Susan Brundage for their insights and for introducing me to many of their patients.

Eileen Perini arranged my visit to the Perini Family Survivors' Center at the Dana-Farber Cancer Institute in Boston. I offer special thanks to her, Dr. Lisa Diller, Amy Grose, and Usha Thakrar.

Dr. Ian Craig and Karen Sugden of the SGDP Center in London tested my DNA for the Resilience Gene, and I thank them for explaining the intricacies of the 5-HTT serotonin transporter.

With resourcefulness and intelligence, my research associate Bridget Samburg pursued survivors around the world and conducted many revealing interviews. She also tackled my obscure interests like the bite force of a *T. rex,* the nutritional content of Taco Bell hot sauce, and the exact name of the blue paint used in handicapped parking spaces. She unearthed treasures every day, and provided no-nonsense feedback, and I am immensely appreciative. I also depended on a number of freelance journalists in faraway places: Alan D'Cruz in Malaysia, Jose Tembe in Mozambique, and Dorit Long in Israel. Thanks to Chuck Lustig of ABC News for introducing me to these fine reporters.

For developing and creating the Survivor Profiler, I'm indebted to Dr. Courtney McCashland and her superb team at TalentMine: Dennison Bhola, Camille Kraeplin, Orville Osbourne, Mary Santoro, and Christian Zoucha.

At Grand Central Publishing, I owe a big thanks to my

dynamic and tenacious editor, Beth de Guzman, and my caring publisher, Jamie Raab. I also extend gratitude to the entire HBGUSA team including Chris Barba, Emi Battaglia, Jimmy Franco, Deb Futter, Laura Jorstad, Kelly Leonard, Alex Logan, Tareth Mitch, Martha Otis, Bruce Paonessa, Jennifer Romanello, Flag Tonuzi, Karen Torres, and Elly Weisenberg. I also wish to thank Maureen Egen, former leader of Warner Books, which became Grand Central Publishing, and David Young, head of HBGUSA.

At William Morris, I applaud and thank Jennifer Rudolph Walsh, who steered boldly and fearlessly from start to finish. My appreciation also goes to Jessica Almon, Claudia Ballard, Tracy Fisher, Eugenie Furniss, Cathryn Summerhayes, and Valerie Suter.

At Creative Artists, many thanks to Alan Berger, Ashley Davis, and Audrey Gordon. For superior counsel, I'm grateful to Craig Jacobson and Gary Schneider.

For support and guidance when this book was barely an idea, I owe much to Irwyn Applebaum, Will Bressman, Barb Burg, Siobhan Darrow, Allen Hammer, Faith Kates, and the incomparable Susan Mercandetti. Carrie Tuhy helped dream up the title and nurtured the project from the start. Joni Evans advised, encouraged, and edited brilliantly at every stage. For brainstorms and incisive manuscript notes and for introducing me to the power of pyschometrics, special thanks to Marcus Buckingham. I also offer appreciation to my friends at ABC's *Good Morning America,* a broadcast that sets the gold standard for survivor stories. In the ABC News family, I want to recognize Jessica Stedman Guff and Robin Roberts, both breast cancer survivors. I also wish to acknowledge James Bogdanoff, a talented senior producer and friend who died of cancer in February 2008.

Bruce Feiler is a fellow traveler, and I benefited every day—often many times a day—from our spirited conversations about life, writing, and publishing. In July 2008, when Bruce was suddenly diagnosed with bone cancer and thrown into the vortex of survivorship, he instantly applied his rare intelligence to "kicking this tumor's butt" (to quote his distinguished oncologist). I felt privileged to be a soldier in his army.

For diving into drafts at different stages, many thanks to Les Firestein and Gwyn Lurie, Ann Hollister and Jonathan Thomas, Mary Jordan, Romi Lassally, Bob Perkins, Julie and Mark Rowen, Andrew Tarsy, Janet Tobias, and Nina Weistein. Dov Seidman, purpose and tenacity personified, shared many meaningful insights. Joannie Kaplan is an empathetic friend who listened early and read often. For enduring countless disquisitions on survival, thanks to Jane Buckingham; Barry Edelstein and Hilit Pace; Mimi Gurbst and Tom Hartfield; Francesca, Sam, and Richard Haass; Lynn Harris; Ann Hollister and Jon Thomas; Linda Rottenberg; David Segal; Katie and Matt Tarses; and Wendy and Steve Trilling. With gusto Matti Leshem grabbed hold of this book and made great improvements. Thanks also to Anthony Goldschmidt, Ron Rogers, and Michael Rosenberg.

For helping to create and launch www.TheSurivorsClub.org, huge thanks to co-founder Gordon Gould, Frank Voci, Brian Weistein at CAA, the creative team at The_Group, and all the angels and supporters and who rallied around the idea.

For taking special care in the life department, thanks to Nisonja McGary, April Rose, and especially Consuelo Peregrina.

On the family front, I wish to acknowledge Arlene, Etta, and Steve Kehela, and Sammy and Gary Mar. Hugs to my sister Liz Sherwood-Randall for her incisive suggestions; Jeff Randall for his surgeon's critique; and young Richard and William Randall for their inquisitiveness. With her exacting attention to every syllable, my mother, Dorothy Sherwood, improved each draft. Her battle with ovarian cancer sparked my interest in survival, as did the untimely death of my father, Richard Sherwood.

He's been gone fifteen years but survives in so many ways. I'm grateful every day to be their son.

In closing, I want to recognize my wife, Karen Kehela Sherwood, a gifted storyteller and editor who elevated every page and endured with equanimity my forays underwater, into the woods, and through the smoke. She contributed in countless ways, but above all, I thank her for her belief and love. In January 2007, she experienced a crushing loss when her father, Edmund Kehela, died after a short struggle with cancer. During his difficult stay in the hospital, the Kehela family rallied together, and I learned so much about many of the strengths explored in these pages.

The Survivors Club is dedicated to my son William Richard Sherwood, a very purposeful little fellow whom we call Will. Someday, when he's old enough to tackle this book, I trust he'll understand what he has taught me about survival. With his vitality and curiosity, he gives me gratitude for every breath and hope for the future. He is the very embodiment of the will to live.

INDEX

ABOUT THE AUTHOR

Ben Sherwood is a bestselling author and award-winning journalist who has worked as executive producer of ABC's *Good Morning America* and senior broadcast producer of *NBC Nightly News.* His acclaimed novel *The Man Who Ate the 747* was a national bestseller and published in thirteen languages, and his most recent novel, *The Death and Life of Charlie St. Cloud,* was an international bestseller. A graduate of Harvard College, he earned degrees in history and economics as a Rhodes Scholar at Oxford University. He lives with his wife and son in Los Angeles.